TOWARD
A CHRISTIAN ETHIC

Toward A Christian Ethic

A RENEWAL IN MORAL THEOLOGY

by

W. H. M. Van Der Marck, O.P.

Translated by
Denis J. Barrett

NEWMAN PRESS
Westminster, Md. New York, N.Y. Glen Rock, N.J.
Amsterdam Toronto

A Newman Press edition, originally published under the title *Het Christus-geheim in de menselijke samenleving,* 1966, by J. J. Romen & Zonen, Roermond, the Netherlands.

Nihil Obstat:
J. Th. C. Arntz, O.P.
Zwolle, March 27, 1967

H.W.M. Bartels, O.P.
Nijmegen, March 28, 1967

Imprimatur:
E.H.J. van Waesberge, O.P.
Prior Prov.
Nijmegen, March 28, 1967

Nihil Obstat:
F.A.A. Schweigman, O.P.
Censor Librorum
Nijmegen, March 30, 1967

Imprimi Permittimus:
M.A.P.J. Oomens
Vicar General
's Hertogenbosch, March 31, 1967

Library of Congress
Catalog Card Number: 67-28699

Published by Newman Press
Editorial Office: 304 W. 58th St., N.Y., N.Y. 10019
Business Office: Westminster, Maryland

Printed and bound in the
United States of America

Contents

Introduction

I

The Question of Renewal in Ethics and Moral Theology

Since World War II, the question of renewal in ethics and moral theology has been more general and more urgent than ever before. This book provides an answer to that question, although it will begin by asking a question about the question itself: What do ethics and morals mean?

Both words are practically synonymous. The way in which they differ has nothing to do with the meaning of the words themselves. "Ethics" comes from the Greek, "morals" from the Latin. Catholics usually choose, or chose, to speak of morals, while others prefer the term "ethics." In a theological or Church context, these latter speak of a Christian or theological ethic, while in similar contexts the current term among Catholics is "moral theology"—a term which is determined very largely, although not wholly, by the title common to certain Catholic manuals in this field since the 17th century.

But these differences between the two terms are incidental. The ambiguity inherent in both terms is of much greater importance.

Ethics or morals can mean, in the first place, that complex of norms and rules which determines the mutual relationships of men in a greater or lesser community, whether or not these norms and rules are given an exact formulation. Drawing an analogy with language, one could call this complex (which includes language itself) the "social language." Every society possesses and needs a language of its own. We may say that each individual is expected to keep to the norms and rules; he is expected to speak the social language of his own community. We can also say that community and communication are possible only because of the existing social language learned by the individuals, so that each is not obliged to invent "expressions" for himself which, in all probability, would not be understood by another who would invent his own, and so on. In this sense we speak of a "professional ethic," while here the word "ethos" is often used also. An ethic, or ethos, is not some particular part of the social language; rather, it is the whole social language considered as the medium of human communication. In this book we reserve the word "ethos" to refer to this reality.

But ethics or morals can also have another meaning and indicate a reflection upon or a science about the significance, various aspects, and implications of human actions, laws, norms, and the like. Ethics in the previous sense, or ethos, is that complex of norms concretely in force in a particular community; it is the social language in actual use. In contrast, ethics, in the second sense, is like the study of linguistics. This is a theoretical study of the fundaments. In the case of ethics, it will aim, first of all, at an insight into all human activity, not being concerned with the particular how, or where, or according to what ethos this activity takes place.

Secondly, this sort of ethics will extend its reflection to the actual ethos of a particular community, seeing in it a specification of the general ethical phenomenon.

On these grounds, in ethics as a science, a distinction is drawn between general or fundamental ethics, on the one hand, and particular or special ethics, on the other. This distinction belongs exclusively to the field of science or reflection, for in the nature of things one can observe only a concrete and determined ethos. It is only by drawing comparisons and by rational analysis that we come to a notion of the universally or fundamentally ethical, just as it is only by comparison of different actual languages or aspects of language that we come to the fundamental significance and structure of language as such.

But let us be clear about the meaning and purpose of a scientific special ethic and let us not be misled by the assertion that reflection extends to the actual ethos as well. A concrete ethic is not, and cannot be, derived by reflection. If a concrete ethic were arrived at in this way, then there would be only one possible concrete ethic, whereas, in fact, it is all too clear that many different social languages do exist (as also do languages for special groups and social dialects).

The fundamental ethic is not the starting point from which one sets out to discover the special ethic; on the contrary, the former is the end at which one arrives by reflecting upon the concrete ethos. In the systematic analysis and discussion which is to follow, the movement is in the opposite direction, not in order to draw special conclusions from general ethical principles, but rather to locate and understand the concrete ethos—every concrete ethos—in the light of fundamental ethical reflection. In this way we can, for example, show how apparently contradictory norms may be found in the same community, such as the prohibition against killing along with capital punishment, or, elsewhere, the same prohibition and the absence of capital punishment, but with permitted abortion, and so on.

There is, therefore, a twofold relationship between fundamental and special ethics. Reflection upon the actual ethos results in a special ethic, and continued reflection leads to the fundamental ethic. Then the movement, as it were, is reversed in order to explain the special ethic and the concrete ethos in the light of the fundamental ethic.

From what we have said, there may arise two questions which are practically identical. One often hears it said that a special ethic is derived from the fundamental. In this way, ethics is taken as a normative science whose task is to prescribe what men have to do and what not to do. Is this not something very different from the explaining and understanding of the special ethic of which we have just been speaking?

Another question which, in fact, differs very little from the above is this: Does what we have said not imply an ethical relativism? In other words, if we cannot derive a special ethic from general ethical principles, but can only "illuminate" and "penetrate" the former by means of these principles, does this not mean that any random special ethic, or any concrete ethos, must be equal in value to any other and prevent our giving preference to any single one? Here, of course, we presuppose that the reader feels a shudder of horror at the mere sight of the word "relativism" or at least that he takes for granted that ethic, or ethos, can in actuality have no plural.

In order to find an answer to these questions, we need, first of all, to remember that the actual community, and thus also the concrete ethos, comes before any reflexive attempt to establish the concrete ethic and, all the more so, comes before any establishing of the fundamental ethic. In other words, neither the concrete ethic nor the fundamental ethic is normative, unless and insofar as they are taken precisely to be the formulation and expression of the ethos proper to the community—an ethos which, besides finding expression in actual behavior and conditioning, is transmitted in definite formulas. Not ethics, but ethos and thus the community itself establish norms. From this it is clearly irrelevant whether or not a special ethic is derived from the fundamental ethic, so that considerations arising from practical prudence in support of that thesis are also irrelevant.

To the question about ethical relativism, the answer can be short. What lies behind the question is, at least to some extent, the preference which one has for one's own ethos above any other. If this ethos, or indeed any ethos, stands revealed as one concrete form of a fundamental datum that is able to take on many forms, the concrete ethos is indeed "relativized." Nevertheless, this does not indicate a value judgment in favor or disfavor of any particular ethos, and because of this the term "ethical relativism" does not apply. Moreover, to arrive at a value judgment is an exceptionally difficult task, at least if one appreciates the complexity of every concrete ethos.

There is another presupposition behind the question about relativism, and it is just as likely to obscure the issue. This is the fear of anarchy, the anarchy that would seem to flow immediately from the relativizing of the concrete ethos. There is no ground for such a fear, for scientific investigation neither confirms nor denies the practical validity of an ethos, although it does affirm the necessity of an enforced, bored ethos.

The ambiguity of the words "ethics" and "morals," which gave rise to the above considerations, is in a certain sense brought out most sharply by the questions of which we have just been speaking. It is characteristic of these questions that, in practice, they identify ethos and ethic, or ethic in the first and second meanings of ethics given above. This leads one to monopolize, as it were, the ethical phenomenon and to identify it with one's own ethos, and so to fall afoul of precisely that ambiguity which can be recognized and avoided only when one appreciates the relative status of one's own ethos. Just how difficult is such appreciation and how easy it is to neglect is shown in a rather striking fashion in the tradition of

Catholic moral theology, although one finds the same thing in other places. Only when one has overcome the limits of this monopoly can one establish the true significance of the very urgent call for renewal in ethics or morals.

The answer which this call expects is thus seen to refer either to the concrete ethos, or to the science of ethics, or, in all probability, to both of these. We can agree that this last is more often than not the case, since the call for renewal in ethics arises out of an unfocused dissatisfaction with the whole actual situation more often than out of a careful analysis of the reasons for this dissatisfaction.

The call for a renewal of morals, in the sense of the concrete ethos, is evident enough in the world of today, considering the growth of international contacts, the increasing intensity of confrontation with other cultures, attitudes and patterns of behavior, and, partly as a consequence of all these, the readjustment of social relationships and expectations. The call for a renewal of morals can, therefore, express a measure of insecurity. The primary answer, then, cannot be in the form of a theoretical revision, but must be a modified, concrete ethos, and this can arise only within the community itself. When we say this, we do not deny that ethical reflection may have a certain influence, but neither do we suppose that such an influence is always necessary.

The same call can also be the expression of dissatisfaction with the reflexive concrete ethic. It should not surprise anyone that this is always some steps behind the factual situation; still, it is right to demand that the reflexive ethic be constantly revised and adjusted.

For us, however, the call for renewal extends only to the reflexive, fundamental ethic. In a sense, all that we have said so far aims to show no more than this: the very presence of an ambiguous call for a renewal of ethics or morals proves, on careful analysis, the need of a basic reappraisal of the ethical phenomenon as such. For the main part at least, our book deals with just this—the fundamental ethic itself.

II

Christian Ethics or Moral Theology

It is not only basic philosophical ethics that is here under reappraisal, but also, and even primarily, the entire theological problematic which has special reference to and arises from ethical situations. This problematic includes, in the first place, all the main questions, although perhaps not all the lesser details, in the Catholic tradition of moral theology, to the extent that these questions must be taken as relevant to our day. Attention will be paid to a number of important questions from the Reformed tradition as well.

III

Aim and Limits

Concerning the aim of this book, there is, in the main, one important observation to make, and this will at the same time clarify an important limit. There is no reason to be silent about the fact that this work on fundamental Christian ethics draws very much indeed on the work of Thomas Aquinas, and especially on his chief work, the *Summa Theologiae.*

Many will not recognize Thomas here, and this is not at all surprising. The genuine Thomas of the *Summa* gives form to his insight, with an exceptional finesse, in hundreds of questions and articles arising from the great multitude of questions, opinions, and discussions in the theology and philosophy of the 13th century. Consequently, it is possible to understand him only if one is familiar with the historical background. This familiarity too often seems missing among those who read, quote, and explain Thomas. They do not really know the historical Thomas of the 13th century, but a Thomas speculatively presented in interpretations dating from the 16th century and later. These interpretations take no account of his real history (we are thinking, for instance, of Billuart, the 18th-century commentator, whose interpretations enjoyed a very wide circulation). We have to realize that, from the 13th to the 16th century, the textbook for and the whole basis of theological instruction was the *Sentences* of Peter Lombard. The gradual replacement of this book by the *Summa* began only toward the earlier part of the 16th century, and already, at that time, a predominantly "speculative" interpretation was far too removed from the actual historical context of the *Summa* to capture the true sense of Thomas' work. It became the fashion to take each question and each article of the *Summa* as a personal discovery and contribution of Thomas without any recognition of the actual historical connections which were the immediate occasion of these articles. This was a regression, a state of affairs parallel to the historical situation before Thomas' time. Worse still, it led to the construction of an amalgam, in the name of Thomas, of all the theological questions of the 13th century, although such an amalgam never did, in fact, exist.

Our purpose now, however, is not to provide an historical interpretation of Thomas' work as such, but a fundamental Christian ethic, the limits of which are given by the author's particular interest in the theology of Thomas Aquinas.

IV

Theology and Rationalism

Precisely because we are now concerned with a Christian or theological ethic, we need to bring out clearly the relevance of yet another limiting factor which is connected with the situation in the whole field of theology. In spite of the worthwhile renewal already achieved, especially since World War II, current theological activity is very largely confined to reasoning, is systematic in character, reflective, and, to this extent, rationalistic, even, and perhaps especially, where we find the continuous tendency to indulge in existential flights.

The first and proper task of theology, however, is the interpretation of scripture: to know the truth proclaimed in the endless variations of scriptural story and metaphor. The great ages of theology were the eras of the Fathers and the Middle Ages, when men were more familiar with scripture than at any other period of history. In those days, metaphor and demetaphorization were the order of the day, long before anyone had dreamed of myth and demythologizing. This was all so strange to the theology of our time that, in 1943, Pius XII in his encyclical *Divino afflante Spiritu* had, so to speak, to grant sanctuary to the "allegorical" exegesis of the Fathers and the Middle Ages to protect it from depreciation and ridicule. Contemporary theology runs, in the main, alongside the exegesis of scripture, which has become a separate, professional, and technical occupation, grouped with patristics and other theological "specialties." It will take quite some time before we see any improvement, since our whole system (of the division of faculties, professorial appointments, specializations, and the like) tends to preserve the status quo. As long as theology itself remains in this state, we shall have to accept the limitations it inevitably imposes in a theological ethic as well.

PART I
Theology and Anthropology

1. Christ

It seems fairly obvious that a discussion on moral theology or Christian ethics should begin by defining a position with regard to the theological character of morals or the Christianity of ethics. In this way it is possible to avoid the confusion and complexity that are usual where the distinction between grace and nature or between redemption and creation is not clearly accounted for and seems to befog every attempt to gain a reasonable view. In this way, moreover, we become obliged to establish our position with regard to the most modern of the variations of this problem: the so-called hermeneutic principle.

When we try to formulate briefly the essential and central content of the Christian revelation and proclamation, we find that we are brought back to the unveiling and expression of what was present, but in a veiled and hidden manner, from the beginning of the world,[1] God's human presence in our own world. We must not think, along with a certain one-sided New Testament bias that one finds in some theologians, that the revelation of Christ in the New Testament is a great reality separate and apart and standing entirely on its own. On the contrary, it is precisely because of the Old Testament that the New Testament is new. It proclaims the perfecting and the fulfilling; it removes the veil; it makes us see what kings and prophets would like to have seen (cf. Lk. 10, 24); it destines something better for us than for the patriarchs (cf. Heb. 11, 40). They knew of God's Word and God's speaking; they knew that the world was the work of his hands, that Yahweh fought for Israel, destroyed her enemies, and brought her victory; they knew that God gave his people food and drink, that he had freed them from slavery, led them through the desert, and brought

[1] Mt. 13, 35; Ps. 78, 2; cf. Col. 1, 26; Eph. 3, 9; 2 Cor. 3, 14; etc.

them into a land that flowed with milk and honey; they knew the prayer, "Thy hands fashioned and made me" (Job 10, 8; Ps. 119, 73).

Farther than this they did not dare to go. The last step, the snatching away of the last veil, was reserved for the New Testament proclamation. Not until this had come did it appear clearly how near God was to men (cf. Phil. 4, 5; Ps. 145, 18). For we must not make any mistake about the meaning of the proclaiming of Christ in the New Testament. Modern biblical exegesis is often enough accused of being risky—and it is, indeed, for anyone who does not, in principle, recognize a limit to the gradual picking and chopping away of myth and historicity and who thinks, nevertheless, that salvation is found, if not in what is mythical, at least in what is historical and thus in what is human and earthly.

However, if we grasp the basic meaning of the Church's traditional conviction about the *truth* of the *whole* of scripture, we see that historical research—as applied to any event or to any person whatever—is an interesting and even, in a certain sense, necessary scientific occupation, but that it does not, and cannot, pass any judgment on the God who appears and who hides in the human reality with which the historical investigation is so eagerly and rightly concerned. If there is any risk and danger, then it is due not to science, but to our everyday humanity and inhumanity, as the description of the Judgment in Matthew 25 bears witness.

We seem to have made quite a jump here, from the proclaiming of Christ to the giving or refusing of something to eat or drink to our fellowman. The precise purpose of this first chapter, however, is to bring out just how essential is the connection between them and how, in a sense, they are the same.

Once again, we must not be mistaken about the sense of the proclamation of Christ in the New Testament. The New Testament is not an historical report about the life of the man Jesus; it is the announcing of God's epiphany, of God's appearing in a human way within the horizon of our own world and of our own daily life. To say this in another way: the proclamation tries simply to open our blind eyes and deaf ears (Is. 35, 5; Mt. 11, 5; Jn. 9; etc.) to the One who in the form of humanity—multiple as it is, interesting and attractive sometimes and sometimes not—is the real Inhabiter of our earth. Or, to say it more precisely, the proclamation reveals that human goodness and goodwill show us God's human face. But it also reveals that for others God's face is not visible unless it becomes visible in us. Salvation and redemption are in human hands, even where salvation and redemption are not even thought or spoken of.

When the New Testament speaks of Christ, it is proclaiming the incarnation of God. Theology has the task of showing that the endless variations of biblical images, metaphors, and allegories are just so many ways of clothing and expressing the central truth and reality of God's human presence in our world. Something that was immediately clear and intelligible to people who lived in familiarity with the Old Testament and with the religious presentations and ideas of the Greek, Egyptian, and Roman worlds at the beginning of our era very often needs closer illustration and explanation in order to get through to us. It is a matter of secondary importance that there are some who, as a result of a particular historical point of departure, speak of myths and of "demythologizing" in this connection, whereas the entire tradition speaks of metaphors and allegories.

Exegesis of scripture must not be allowed to become a (profane) history, sociology, anthropology, or ethics. It is to be hoped that the period is behind us when, with good intentions but little intelligence, it was the fashion to *proclaim* that the story of Adam and Eve did not really happen, that Jonah did not really remain inside the whale, and that the account of Noah, his ark, and the flood was just another story.

Perhaps it is fortunate that we did pass through this period of nibbling away at the Old Testament and at some parts of the New, because it allowed our excitement of discovery to cool off and gave us time really to appreciate what we mean when we say that the *whole* of scripture is *true,* before an historical, critical technique—quite inadequate to Revelation—had brought us to a nihilistic result. Perhaps it was necessary for us to speculate about the psychological ego of Christ and about his gradual appreciation of his own divinity and of his divine mission before we could realize that, in such psychological speculation, we were on a wrong road. Perhaps it was necessary for us to appreciate the fanatical eagerness with which some liturgists propounded Easter as the greatest feast of the Church year and deprecated the romanticism of Christmas before we became able once more to perceive that such great portions of Eastern and Western tradition were able to maintain a genuine "epiphany" only because birth and death and resurrection, each in its own way, point to the one mystery of God's humanity (cf. Tit. 3, 4). For this is a mystery which cannot be confined within any one single expression, as the fundamentalists would have it. It is a mystery which one can accept or reject, not without happy consequences to the acceptance, or confusing and unhappy ones to the rejection. Perhaps it was necessary for us to have experienced a period of post-modernist fundamentalism, during which there was a gradual preparation of thought and energy, leading at this moment to a "reform in head and members," a reform accepted with gratitude and served with responsibility.

The reform concerns us here only to the extent that it involves a revision and enrichment of our theological and scientific thinking, and in this the key position is held by christology. People are calling, quite rightly, for a christocentric theology and a christocentric moral. Not quite so rightly, as I see it, the same people come out in opposition to traditional dogmatics, which, in its teaching about God, actually contains more of christology than a too one-sided interest in the historical Jesus was ever capable of achieving, or in opposition to the current moral teaching which, in its fundamental adherence to natural law, took God's incarnation much more *au sérieux* than any gaily gliding enthusiasm for the bible could imagine.

Christ's place in morals—whether in Thomas' work or elsewhere—is not determined by quotations from scripture, but by God's incarnation itself. When scripture speaks of Christ, it is speaking of him through whom all things were made and without whom nothing was made (cf. Jn. 1, 3), and in whom all things have their being (cf. Col. 1, 17). Only the man who is quite unfamiliar with all that the Old Testament has to say about God's Word and God's wisdom, or who is unaware of the error of Nestorianism, can miss the divine breadth of this Christ. Apart from him there is no humanity, and all humanity there can possibly be exists in him. In other words, there is nothing human that does not show forth the face of God, and it is the face of God himself that becomes visible to us in all that is human. Creation is seen to be incarnation—and redemption—and for this reason traditional dogmatics are "'christological." Nature is seen to be grace (we shall return to this later), and for this reason the traditional morals are "christocentric."

Ethics cannot be other than Christian ethics, not because of nature or humanity, but in virtue of God's incarnation and revelation of himself in humanity. Thus we speak of a Christian ethic not in order to indicate a conviction that there might be a non-Christian ethic as well, but simply in order to say that the human ethic is, in fact, Christian. This means, moreover, that our work can be a genuine Christian ethic only if we constantly and logically acknowledge human autonomy and allow to it its proper function.

This autonomy—the key concept in Schema 13 of Vatican Council II—is an essential presupposition of the Christian ethic. Without it, no Christian ethic is possible. It is but another aspect of God's incarnation.

God and man are not competitors. Man does not exist in spite of God, but thanks to God, and man is not deprived of anything by the fact that he gives form to the humanity of God. On the contrary. How could it be otherwise, and what would man otherwise be? Man is not free in spite of God, but thanks to God. Human autonomy does not in any way diminish God, but is, on the contrary, a divine gift. Human activity, intersubjectivity, goodness, and benevolence do not diminish the incarnation and redemption; on the contrary, human reality is the form of incarnation and redemption brought about and taken on by God himself.

To identify God with evolving humanity would be a fatal mistake. But if we wish to continue speaking of both the transcendence of God and his immanence, we must beware not only of a pantheistic presentation, but also of a deistic presentation which imagines that it has saved and preserved God when, in fact, it has only denied his immanence and therefore denied his incarnation. And this is what usually happens where we find a latent presupposition that divinity and humanity are in competition dominating man's reflection about God and the divine or grace and the supernatural.

In other words, just as we, as *viatores*—people of this world—cannot meet any other God than the God who is with us in the humanity of our own world, so, too, our theology cannot be other than christology, because it is, in fact, the theology of the immanent God, of the Transcendent who has become man.

So, too, we see that the famous old problem of the motives for the incarnation rests on dubious presuppositions—in the last analysis, on a (profane) historical interpretation of scripture. The parts in this discussion are taken in the name both of the so-called Thomistic thesis that God would not have become man if Adam had not sinned, and of the Scotistic thesis that God would have become man anyway.

Now, in the first place, the notion of an Adam who did not sin is a contradiction in terms, in the sense that man, in terms of the actual economy of salvation established in creation and the incarnation, is, by definition, he who sits in darkness and the shadow of death insofar as he is captive within the horizons of mere humanity and lacks any perspective on God's presence in his world. This is precisely what the salvationless state, or sin, is. In this sense, Adam is quite impossible unless he be either in sin or redeemed (or "engraced").

Secondly, the parties in this dispute speak of God's becoming man as though it were an historical fact that took place at an historical moment after an historical fall into sin, and, to pinpoint it more closely, about the year 1 by the current calendar. However, God's becoming man and his presence in humanity is not a fact that can be determined by historical investigation, but it is a salvational event from the very foundation of the world itself, a saving truth and reality which man can recognize only after God has opened his eyes to it.

In other words, the dispute about the motives for the incarnation is yet another of the many examples, not of theology, but of a pre-theological manipulation of texts and allegories from scripture. This is merely a preliminary to theology. The discovery of the fact that some texts, taken together, "do not jell" and that others again are in harmony is an elementary phase in the evolution of a genuine theology, which consists in so laying open the deepest sense of scripture that the apparent discrepancies among the most diverse allegories and modes of expression are removed.

For a theology which was, in fact, a fundamentalist manipulation of texts, or a pre-theology, rather than a genuine theology, the discovery and formulation of the "hermeneutic principle" marked the transition to a genuinely theological phase. Of course, the hermeneutic principle can be formulated in many ways. The most general and, in a sense, the most obvious formulation could be this: the primary principle to maintain in all investigation is the *truth* of the *whole* of scripture. To be less formal and more to the practical point, one would have to say that the truth of the whole of scripture relates to *the* mystery—God's presence and his visible appearance in humanity, within our own world, which is not only the work of his hands, but the shape of his unsearchable goodness, love, humanity, and benevolence.

Our understanding of the revelation of Christ is helped equally when we see it in the light of the Church and, particularly, of the catholicity of the Church.

We are accustomed to use the word "catholic" in a sociological sense, as the label of a particular group, or as a qualification of the views and structures of that group. This usage, however, must not prevent us from seeking the original and fundamental sense of the term "catholic"—for example, the sense in which it is used in the Creed.

If we can appreciate this word not merely in terms of its Greek meaning, but much rather as a formulation of the essential content of revelation, we see that the term "catholic" ("universal") expresses the conviction that God's visible appearance in humanity—in *all* humanity—brings salvation and its opposite within the reach of any man whatever in any condition whatever. "Belief in the Catholic Church," therefore, does not primarily mean a conviction that a particular social organization rightly directs its preaching of salvation to all men of all times in all places. In fact, the primary meaning is rather the positive and actual recognition of the God who is making himself visible to us in every man, as gift and as demand at the same time.

Catholicity is not a human claim, but rather the actual universality of God's visible presence in mankind, offered to us and to be accepted by us, without any reservations on grounds of color or culture or the ethos in which his humanity takes shape. Describing catholicity in this way, Vatican Council II, in the Constitution on the Church, expressly declared that the Church neither diminishes nor opposes, nor in any way can diminish or oppose, the values proper to this world or the morals of any community or society whatever.[2] On the contrary, the Church acknowledges the proper legal-

[2] *The Constitution on the Church of Vatican Council II* (New York: Paulist Press, 1965). Note particularly n. 13: ". . . the Church . . .

ity or autonomy of civil society[3] and the earthly task of its citizens. The Church has the responsibility of bringing out clearly the deepest sense of this earthly task, which, in fact, is seen to be a divine and saving task,[4] because God wants to show himself to us in our fellowmen and to our fellowmen in us.[5]

In this Church, and in her catholicity, the necessity of ecumenism is obvious, again not as a human endeavor, but as something implicit in the mystery of Christ—God's having become man in all humanity.

However much God's human presence in the world is a fact, it remains, nevertheless, a veiled presence and can even be threatened by inhumanity or by arrested or inhibited humanity. This veiled and hidden presence comes to light and, in a certain sense, becomes a real presence for us only in its explicit revelation, through the confession of faith, through proclamation, preaching, and prayer: in short, through sacramental expression and positive enactment which not only contain an explicitly worded acknowledgment of God's engracing presence, but also bring about an obvious unity and solidarity among all who share in his presence.

Thus, Church, proclamation, and sacramentality all coincide in one sense, and in another do not. "He who loves is born of God,

---

does . . . foster and adopt, insofar as they are good, the ability, riches and customs of each people. . . . In virtue of this catholicity each individual part contributes through its special gifts to the good of the other parts and of the whole Church." See also n. 17: "Through her work, whatever good is in the minds and hearts of men, whatever good lies latent in religious practices and cultures of diverse peoples, is not only saved from destruction but is also cleansed, raised up and perfected unto the glory of God. . . ." In n. 32 it is stated: "There is, therefore, in Christ and in the Church no inequality on the basis of race or nationality, social condition or sex. . . ."

[3] *Ibid.*, n. 36: ". . . It must be admitted that the temporal sphere is governed by its own principles, since it is rightly concerned with the interests of this world. . . ."

[4] *Ibid.*, n. 48: ". . . in which we learn the meaning of our terrestrial life through our faith, while we perform with hope in the future the work committed to us in this world by the Father, and thus work out our salvation. . . ."

[5] *Ibid.*, n. 50: "In the lives of those who, sharing in our humanity, are, however, more perfectly transformed into the image of Christ. . . ."

and knows God," writes St. John.[6] The primary form of religion can be only that in which God is served where he primarily and normally presents himself—in other men. However much common humanity, religion, and faith may be distinguished, they do, in fact, coincide. It is God's human presence in the world (we are still speaking of the mystery of Christ) that makes the second commandment like to the first commandment, and which shows that the dilemma of horizontalism versus verticalism is really a false dilemma. For our fellowman is not a mere chance or coincidence; he is the manner in which God is pleased to appear to us and to be called "God-with-us." There are some who call this actual presence and recognition of God a religionless Christianity, but the term is not well-chosen because, in fact, no religion can be imagined that is more fundamental than this. Furthermore, how could there be any community at all that is not community with God, which is to say, that is not Church?

At the same time, however, it is true that Church, salvation and community with God are real for men and become redemption and liberation only within the context of their being explicitly proclaimed and made sacramental. Understood in this way, this Church cannot be any self-sufficient, self-satisfied club or coterie isolated in herself, because in the very nature of things she must know herself to be at the service of each and every man, fundamentally, toward the constructive progress of common humanity and, consequently, in the proclamation not of a manual of dogma or a moral code, but of the mystery of Christ, of God's human presence in our own world.

In all the above we have attempted only to bring the mystery of Christ to the forefront of our stage by approaching it from different angles. This basic theological introduction was needed because it is only in this light that we can see how we have to go about setting up a moral theology, a Christian or theological ethic. We have to set about our work in an autonomous, human way; this is not only justifiable, but it is shown to be strictly necessary by the very nature of the Christian revelation. Scripture does not

[6] 1 Jn. 4, 7. Note also the antiphon in the *Mandatum* rite of Holy Thursday: *"Ubi caritas et amor, ibi Deus est."* Cf. also St. Thomas, *Summa Theologiae,* II–II, 108, 1 ad 3: ". . . *qui ex amore bonum operantur . . . soli proprie ad Evangelium pertinent."*

give us an ethic, but only the certainty that the human ethos itself signifies salvation or the lack of it. Even in the Church's tradition we may not in the first instance look for more than an ever-repeated witness to the real solidarity of humanity and salvation.

This does not mean that we are simply abandoned to our own inventiveness (or lack of it). The catholicity, or universality, of God's offer of salvation-in-humanity certainly forbids us to remain confined within any one *a priori* standpoint, but it as certainly does not permit us the shortsightedness, the self-sufficiency, and the laziness of neglecting and undervaluing the wisdom and experience of others, present or past.

On exactly the same grounds, we are not confined to a circle of those who think alike. Schema 13[7] directs Christians to man and to the world, requiring that men cooperate with *all* their fellowmen for the carrying out of their earthly task and, in particular, for the solving of ethical problems. This is not merely a diplomatic gesture, but, once again, an expression of the very nature of the Christian proclamation and of catholicity.

[7] *The Pastoral Constitution on the Church in the Modern World,* n. 16: "Through loyalty to conscience Christians are joined to other men in the search for truth and for the right solution to so many moral problems which arise both in the life of individuals and from social relationships."

2. Man

Just as a Christian or theological ethic is based on particular presuppositions about God, the mystery of Christ, and theology, an ethic must be based on a particular view of man, a particular anthropology. It seems quite obvious to me, therefore, that the next point for discussion must be anthropology.

It is a very noticeable and, to my mind, a welcome development that the concern with the understanding of "being" in classical metaphysics has, in present-day philosophy, given place to a great interest in man and anthropology. It can hardly be doubted that the progress of such sciences as psychology and sociology is symptom and, at the same time, cause of this.

If there is one thing that may be called fundamental and central in anthropology, it is intersubjectivity, communication, love, human solidarity, justice, or whatever other name one chooses to call it. For, in the last analysis, all these terms express one and the same human reality. The reality is this: Not only is it not good for man to be alone, but man in his very essence is not an isolated individual; the whole of his nature directs him toward community and fits him for it.

In the first instance, man cannot be entirely distinguished from all those phenomena of constant action and interaction that make up the normal pattern of the whole world of matter. In a certain sense, man is simply absorbed into the ever-advancing process of reciprocal influences unfolding in matter. Man in his totality is a material, physiological, biochemical reality, subject to all the laws of matter.

However, we must at the same time note that this same man possesses the unique possibilities of understanding, love, goodness, attention, benevolence, tact, and creativity, and it is precisely these that mark out this particular material and physiological being as man. In other words, the whole of that bodily reality which is man exists in function of that unique intersubjectivity, community, and communication which is the peculiar possibility and task of man. We can say the same thing in yet another way. Corporeity and intersubjectivity—or, if you will, body and soul—indicate one and the same material reality and, at the same time, two realities that in a sense are entirely different. Now we are certainly not agreeing with any dualism which takes body and soul, or materiality and intersubjectivity, as two materially distinct realities, of which one, the soul, resides and dwells in some mysterious way in the other, so that at death it can be released from the shell of the body to fly to its heaven on high. In order to realize how widespread and unquestioningly accepted such a view is, one has only to reflect a moment on the way in which we have learned or tend to think of body and soul. If this were not enough, sufficient proof of the generality and unquestioned acceptance of this way of thinking can be found in one of the most passionate disputes in the history of theology and human thought, the controversy *"de unitate formae [humanae]"* (on the unity of the human soul) that arose after the death of Thomas in the last quarter of the 13th century. In all probability, there is no aspect of Thomas' teaching that became the object of such unfounded lip-service on the one side, and of such exceptionally bitter criticism on the other, than his view on corporeity and intersubjectivity, body and soul.

In addition to all this is the fact that conventions of language alone are scarcely or not at all able to avoid the likelihood of a dualistic interpretation. If it is said that the human spirit renders itself incarnate in matter, the expression itself—not entirely without reason—gives a fairly strong impression of dualism. Nevertheless, it is possible that such a formulation could arise from a manner of thinking and a point of view which were not dualistic. Even when it is said, as we noted above, that corporeity exists entirely in function of intersubjectivity, it is often necessary explicitly to state that this formulation must not in any way be understood in a dualistic sense. For we are so accustomed to live with "things" that are materially distinct, and the conventions of our language are based to such an extent on this experience, that we find it necessary to exercise the greatest care and circumspection when we want to enter upon more fundamental considerations.

Thus, corporeity and intersubjectivity, or body and soul, are two terms that indicate materially the same reality—man—but nonetheless, in a certain sense, refer to two realities that are entirely distinct. For body, or corporeity, indicates man as a somewhat insignificant part of material reality and of the material process of action and interaction. Over against this, soul, or intersubjectivity, characterizes man in his proper nature and in his unique difference from all the rest, with his possibilities of understanding, love, and so on.

A special difficulty again presents itself here in the particular background of these terms. Body, or corporeity, points emphatically to the material thing, in a certain sense isolated from all other things, whereas in this context we want the term to bring out precisely the interaction, or, if you will, the dynamism that is proper to all material reality. This is not, of course, to suggest that interaction is something that belongs only in the human sphere, but it is intended to bring out that special interaction which is specifically human and which we call understanding, love, goodness, and the like.

We hardly need explain how the heritage of the word "soul" has loaded its meaning. The "shining" or "stained" soul that we remember from our catechism lessons in childhood is only one of the remarkable fantasies connected with it. It is significant that this use practically neglected the factor of intersubjectivity, making the role of God and the devil all the more dominant. At first sight, it seems as if there is scarcely any point in so radically spiritualized and supernaturalized a concept upon which we could begin to base a discussion that would be anthropologically sound. Nevertheless, our intersubjectivity is bound up with the common conventions of our language and also with a great deal of our history, and this means that we are obliged to use these and similar words, even if it also means that we shall have to eschew certain undesirable aspects connected with and often inherent in the use of these words.

Thus we accept the words body and soul—that is to say, corporeity and intersubjectivity. Perhaps we can add, in the context of the purification we have undertaken, and preferably now than later, that they are related to each other as means to end. "Means" and "end," let us immediately note, are not to be understood in terms of their conventional usage, but in terms of the meaning of body and soul, of corporeity and intersubjectivity, as we have just explained them.

Means and end are not terms for two materially distinct realities, but for two formally distinct aspects of a reality that, materially and practically, is one. For example, man is one reality in whom corporeity and intersubjectivity, or body and soul, are related as means and end. It was in this sense that we said earlier that the entire material, physiological, biochemical reality, which in a certain sense man is, exists in function of intersubjectivity, which alone is really and fully man.

For greater clarity, we may give just one example. A man's walking can be described in terms of muscular movement, biochemical reaction, brain-cell operation, the consumption of energy, and so on, and all this is in the sphere of man's corporeity. But as a human reality—that is to say, in terms of intersubjectivity—this physical "walking" can be practically anything: going for a stroll, going to the office, serving coffee, modeling a new fashion, trying out new shoes, departing in anger, and so on. Something that, physiologically, is a movement of the lips can, humanly speaking, be talking or eating or sipping or whispering or kissing or anything else that lips can do.

The corporeity of man is intersubjectivity; his body is his soul—even if this last statement sounds strange. If we are to use the words "means" and "end" for the right reasons, we must see the body, or corporeity, of man as means, and soul, or intersubjectivity, as man's end or goal. It is impossible to say that corporeity and intersubjectivity are two things separate and distinct from one another in time and place, for, on the contrary, practically and materially they coincide; and it is equally impossible to say that a means and its end are two things separated spatially and in time, however much our ordinary use of language or our usual manner of picturing these things in the mind may incline us to do so.

The examples given, of walking and of moving the lips, were taken from the sphere of human actions for reasons that speak for themselves. For activity in one form or another belongs so properly to man that it is almost impossible to present a notion of man as such, unless he be presented as active in some way. The description or characterization of man begins spontaneously with his activity. When we wanted to indicate what was proper to man in connection with the interaction of material beings, we chose words such as understanding, love, and so on.

In speaking in this way, we need to realize that we are not speaking of things that are accidental to man, but of man as such. More precisely, we should not make the distinction in this way at all. When we speak of accidents, we are speaking of man strictly as such, at least as long as we do not have the incorrect notion that accidents are the same sort of thing as the baubles on the Christmas tree which may be removed and still leave the bare substance of the tree. An accidental difference is such that it gives one the impression of being confronted with an entirely different reality, while a "substantial" difference may conceal itself under accidental sameness.

Proceeding from man's activity and understanding him in his activity, we have to say of man that he exists, as it were, in a web of innumerable contacts, some supple and elastic, others hard and brittle, that bind him to other people, whether to a few or to many or even to countless numbers. The web makes a kind of field of force in which every movement that a man makes causes changes, new tensions, releases, and sometimes even ruptures.

The very essence and nature of man is intersubjectivity. It is not something he can choose to avoid or escape. On the contrary, from the very first moment of his life, it is coexistent with his humanity. All that he is, he is thanks to intersubjectivity, or whatever else one wishes to call it: love, goodness, humanity, benevolence, or the like. Physiologically, man from the first moment of the embryonic union of sperm and ovum is the fruit of material interaction, but, precisely as man, he is the fruit of love and intersubjectivity.

In his turn, the new man takes his place in this wonderful process of intersubjectivity, having been made ready for it by other men. The process itself poses the questions: What is man? What am I? In a certain sense, I am not, or am hardly, a whole being in my own right. I am, nonetheless, something wholly my own. I am, in a certain sense, the product of existing with and encountering numberless others, some more than others; the product of goodness and trust, but also of distrust and suspicion; the product of resoluteness and honesty, but also of indecision and intrigue;

the product of balance and realism, but also of instability and dreamy idealism; the product of openness and ambition, but also of shyness and suppressed lust for power. And so we could go on.

If something on the material or physiological plane is untrue, on the human plane also it is pure illusion. Only a completely unreal abstraction could allow us to conceive of man as an independent, self-sufficient whole. Man is not such. Just as he is materially in constant interaction with his surroundings, so, precisely as man, he is essentially and constantly intersubjective. He is neither simply the passive product of interaction nor wholly the fruit of intersubjectivity, for he himself takes an active part, and just as what he is is partly dependent on others, so what others are is partly dependent on him.

Therefore, if we want to speak of ethics—that is, if we want to consider human action and all that is connected with it and inherent in it—a realistic, anthropological pre-reflection is of fundamental importance. For it is not possible to speak of human activity without definite presuppositions regarding man and his activity, whatever kind of presuppositions and however unspoken and probably vague they may be. The architect is responsible not only for the construction above ground level, but also for the foundation on which the construction rests. However hidden the foundation may remain, however little attention it may draw, and however much it may lack the spectacular lines, surfaces, and colors of the superstructure, it is no less indispensable, and it is better to busy ourselves about the foundation before cracks and settling in the superstructure begin to draw attention to it.

This comparison, obviously, is only partially applicable. The situation in anthropology and ethics is not quite the same as with foundations and superstructure or axioms and derived theses in mathematics, for example. Anthropology is not a *Vorverständnis,* a primary principle of understanding to which we must all *first* subscribe before we can discuss the conclusions. We met something of the same sort in the case of the prerequisites to a genuine theology of which we spoke in the previous chapter. Ethics, or the Christian ethic, is a continuous consideration in which the *Vorverständnis* itself is taken into and remains within the discussion. In ethics we are not concerned with principles from which we can draw neat little conclusions, but with a consideration and delineation of human reality in all its diverse aspects and implications, having constant reference to the reality itself. The so-called foundation, or *Vorverständnis,* remains constantly an object of discussion, criticism, and refinement. Thus, although it may appear logical and of principal importance first to reach agreement on the *Vorverständnis* before going on to further considerations, the fact is that the only way to achieve any insight into matters in this science is to get to work on those of its many aspects and implications that first present themselves to the investigator, weighing carefully each thing along the way.

This preliminary anthropological discussion must, therefore, in a certain sense be verified in the ethical discussion that is to follow, just as theological presuppositions must be kept in mind when we are speaking of the different aspects to which ethics, precisely as Christian ethics, will direct us, partly as the result of certain historical and traditional points of view, formulas and data.

If, therefore, we take corporeity and intersubjectivity as the anthropological starting point of our ethical discussion, the reason is, first, to place systematically things that are, in fact and in their genesis, the goal and the result of investigating concrete moral or ethical problems. These problems compel us to establish, or to re-establish, the fundamental point of departure or, rather, the implicit, unexpressed, presuppositions. The problem of capital punishment, confronted with killing which is evil in itself, the problem of birth regulation, confronted with sterilization which is evil in itself, the problem of abortion, artificial insemination, nuclear weapons, and so on, each with supporters and antagonists—these all compel us to a reconsideration of the "principles" of the natural law, of what is actually good and what is evil. All this leads us finally to a consideration of human existence itself.

We shall discover, through this approach, how easily man is often identified with the kind of being that exists in itself, alone and unrelated, a whole unto himself. In short, man is often taken for and identified with an abstraction, which, in actuality, he is not. Person *is* community. This was the cry that arose, shortly after World War II, in circles involved in personalism, although many years had to pass before their conviction could gain any significant ground. But now this formulation is able to serve as a brief summary of the whole of anthropology.

Conclusion

We began our consideration of moral theology, the theological or Christian ethic, with two chapters: one on Christ, the other on man.

The essence and center of the mystery of Christ, as announced in the Old and New Testaments, is the epiphany of God, his becoming man, his human manifestation among us in our own, everyday, human world. This is a manifestation that does not compete with what is human, but, on the contrary, unveils and brings into the light the deepest nature and content of man and all that is human, then re-creates it all and brings forth the new earth.

What scripture and revelation bring us is not primarily an ethic. They do make it clear to us, however, that ethos and ethics are man's concern, just as man's autonomy is man's own concern, and that an autonomous, human ethic is part of the task entrusted to us in the context of God's plan of salvation.

Anthropology lies at the basis of this ethic. It was therefore necessary, before beginning to consider the nature and the different aspects of human activity, to think for a while of man himself. Man is not a monad, an independent being separate and apart, sufficient unto himself. Man is not alone, either as to his corporeity or his intersubjectivity, his love, the humanity he shares with others—all of which constitute the peculiar nature of man and, therefore, his life task. Man's corporeity is intended as, and directed toward, intersubjectivity; it exists entirely, therefore, to subserve this intersubjectivity. Ethics will reveal what the implications of all this are.

PART II
Fundamental Christian Ethics

1. The Meaning of Human Life

During the whole of our known history, the question about the meaning of human life has, understandably, been the essential question in ethics. Systematic ethics, after the example of Aristotle, often began with a consideration of this question.

Christian tradition imprinted an entirely unique stamp on this ethical question. What had been a problem for the non-Christian was not a problem for the Christian, for he knew from revelation that God was the goal and the destiny of man. When the Christian considered the *eudemonia* about which the Greeks puzzled, and the *felicitas* for which the Latins searched, he found his answer in God—often in terms of *beatitudo,* for this word was suggested by many sources, and chiefly by the Sermon on the Mount.

Christians did not altogether fail to see that this answer did not solve the philosophical and ethical question about the earthly destiny of man—one has only to think of Augustine and of his disputes with the philosophers of his time—but in the West this was probably never so clearly seen as in the Middle Ages.

The theological conflict of the 12th century—between the party of the traditional, sometimes traditionalistic, theology of the *Sentences* (with Bernard, the schools of Laon and St. Victor, and others), and the party of the more rational and critical, sometimes rationalistic, theology (with Abelard, Gilbert Porreta and their followers, and others)—was continued with many refinements into the 13th century. During that century, however, a whole new development emerged. The rediscovery of Aristotle's works and those of various commentators, as well as the great interest shown in them, stimulated to no small extent by ecclesiastical prohibitions, led the *facultas artium* of the University of Paris gradually to assume the character of a faculty of philosophy (Siger of Brabant). Theologians were faced with a number of critical questions and problems which the traditional theology of the *Sentences* could not solve. Confrontation with an autonomous philosophy and ethic forces theology to undertake an extremely critical enquiry into its own content and method. William of Auxerre, Philip the Chancellor, the *Summa fratris Alexandri,* and Albert all made outstanding contributions to this investigation, but it can be said, without injustice to anyone, that its acme was not reached until Thomas wrote his *Summa*—a claim which later became uncontested.

For the first time in a systematic Christian ethic, and influenced in part by Aristotle's *Ethics,* Thomas, in the *Summa,* gave first place in Christian ethics to the question of the meaning of human life. It was only in the 16th century that his treatise *"De ultimo fine et de beatitudine"* ("On the Final End and Divine Destiny of Man") seemed at last destined to become more influential and to receive wider circulation. At the same time, however, its significance was obscured and limited because it was given a sectarian flavor, if one may use the expression, by the disputants in the schools who employed it in arguing about the *"constitutivum formale beatitudinis"* and similar peripheral questions. This was also the period of the Council of Trent, and Trent had revised the curriculum for the education of priests. The theological teaching of the time was directing moral theology into a more and more practical framework, so that "theoretical" questions, such as that on the meaning of human life, came to be ignored altogether.

The ethics of the 17th and 18th centuries, not altogether unjustifiable in this respect, did not begin to be corrected until the 20th century, often explicitly in conjunction with the opening articles of Thomas' theological ethics. Meanwhile, the depth of insight shown in these revised manuals was not always conspicuously great.

This means, in a sense, that we stand once more before the same question. Current convictions about the absurdity of human existence, or about the inauthenticity of the appeal to God and of all that is religious, put obstacles in the way of a premature scriptural and Christian witness to the meaning of human life. Marxism and existentialism have brought up questions about man and about the earth which take away from us the right to speak immediately and exclusively about God and about heaven. On the other hand, Christian tradition gives us a fundamental guarantee of the authenticity of the human answer, an answer that can be given before any mention of salvation or the lack of it need be made; what is more, an autonomous, human answer will be able to purify and strengthen our Christian conviction.

Up to now we have deliberately spoken of the "meaning of human life," and not of "happiness," even though this latter term may well have been suggested by certain Greek and Latin expressions (Gr., *eudemonia;* L., *felicitas, beatitudo*) and by their customary English equivalents. The entire human problem with which we are now concerned is already contained in the difference between these two terms.

Let us begin at the beginning—that is, with the point with which Thomas opens his theological ethic, and which, in one way or another, has continued to play a role in all considerations right up to those of the present. The point in question is "action in view of an end"—*"propter finem agere."*

When we use this expression and speak of finality, we think, just as when we use such words as intention, goal, and the like, of a fully conscious and carefully deliberated exertion in order to achieve a particular aim (for example, the Peace Corps program of the United States). Up to a certain point, this is correct.

However, when we examine more closely and precisely the human reality that a very long tradition has characterized in the formula *"propter finem agere,"* it becomes clear that the simple translation, "to act in view of an end," is very much less exact and adequate than we at first imagined. The human reality, in a certain sense, is not primarily action that normally would be very consciously directed toward a goal. If human action were normally like this, it would prove very exhausting. The human reality is primarily rather this: the actual implications and consequences of our actions, for our fellowmen and for ourselves, can always be formulated in terms of intersubjectivity. In other words, *"propter finem agere"* is not primarily a psychological formula, but an anthropological one, and it indicates the *factual* intersubjectivity of human action even before there is any question of a possible, conscious, mental reflection on the actual results of the action. Modern psychology, with its discussions on the unconscious, on

Fehlleistungen (unperceived ulterior motives), and so on, has, to my way of thinking, helped to some extent to bring to light once more the original, anthropological sense of the expression *"propter finem agere."*

The fact that *"propter finem agere"* indicates, in the first place, the factual and, in a sense, necessary intersubjectivity inherent in every human action does not in any way mean that it does not also indicate reflexive consciousness or concurrent or premeditated intention of the actual results of the action.

There is a tendency here to object that genuine human action, on the contrary, must contain precisely this element of consciousness; but this tendency deserves to be kept under careful control. For there is danger that one may demand too much reflection and that one may lay too heavy a stress on intellectual consciousness, calling for something far more burdensome and more one-sided than is needed for genuine human action. We must return to this at a later stage. For the moment it is enough to say that, in general, man is rightly looked upon in terms of the actual results of his action, and he is rightly held responsible for them. For man, as we say, knows, or is supposed to know, what he is doing—in contradistinction, for instance, to an animal. Thus, in a certain sense, *"propter finem agere"* lays the main stress on *"agere."* For, indeed, an actual result belongs inherently in every action and in every activity. What is typical of man and belongs uniquely to him and to his activity is perhaps not so much the fact that he can "see beyond" what he is doing as the fact that in his action he takes part, in some way, in intersubjectivity, that he himself is actively intersubjective.

It will be clear that in this analysis of human action, of the *"propter finem agere,"* we have already covered a good deal of ground in the direction of an answer to the question about the meaning of human life. True, up to now we have spoken only of *"propter finem agere,"* although the traditional title of this chapter, in Thomas' work, is *"De ultimo fine et de beatitudine."* Nevertheless, this *ultimate* goal of human existence in the world is, in fact, under discussion from the moment one begins to speak of the intersubjectivity implied in all genuine human action. Undoubtedly, human action implies much besides intersubjectivity. There are implications on the level of matter and control of environment, and these, as factual results of human action, are equally inherent in it.

Nevertheless, the strictly human significance and implication of all activity is intersubjectivity, and, humanly speaking, this is at the same time the ultimate and most fundamental thing that can be said of human action. The ultimate aim of man is none other than intersubjectivity, communication, community, a share in common humanity, love, justice—or whatever more or less historically conditioned or even loaded term one wishes to use.

In a certain sense we have just explained why the title we chose for this first chapter is "The Meaning of Human Life," and not "The Happiness of Man." No one will find it strange that every man wishes to be happy. Neither will anyone disapprove of that longing in another or find it suspect; rather, it will be accepted as the most obvious and natural thing in the world.

But none of this argues against the fact that the longing for happiness is not a suitable starting point of a systematic ethic. Not only do the longings and desires of men often take very different directions, but long human experience and wisdom teach, above all, that the less a man worries about his own happiness, the happier he becomes. This follows in the nature of things from man's essential intersubjectivity. It is his nature and his task to take part in and to make his contribution to intersubjectivity. For every man this is, in fact, the sense, the necessary sense, of his life, the common basis of every ethos, and, consequently, of every ethic.

Nothing is clearer, as experience bears witness, than the fact that man will find the happiness and joy he so rightly desires by taking his share in intersubjectivity and contributing his part to it. But happiness and joy are the product, the consequence, and the fulfillment of the meaning of life; they are the return gift which one receives because of and in answer to an unworried care for intersubjectivity, for others, and for the community.

In a certain sense, therefore, there is something remarkable, a turning upside-down of human reality, in the question sometimes anxiously asked: Is this person really happy? How can he become happy? For all human happiness is essentially connected with fulfilling the meaning and task of human intersubjectivity; it is, as it were, the obverse side of this fulfillment. Unless, of course, the question about someone's happiness proceeds from a genuine solicitude for the possibilities of intersubjective action and communication upon which his happiness depends.

The fact that the meaning of human life, and thus intersubjectivity, is taken as the starting point of ethics—and that happiness is not taken as its starting point—does not therefore mean that happiness is ethically irrelevant or even suspect, as it is in the ethics of Kant for whom the bare fulfillment of duty was the ethical ideal. But it does mean that one is able (with Kant) to break through and avoid the egocentricism of a eudemonistic or "happiness" ethic (in which everything seems to exist in function of the ego) without, on the other hand, losing sight of a genuine solicitude for man's happiness.

The Christian proclamation has something to say about this human reality. Intersubjectivity, as a human gift and task, is in reality something more than a merely human thing; it is the form of God's presence in the world. Thus, in a certain sense, the meaning of human life remains the same as it would be without any Christian proclamation and as it was before the Christian proclamation; but this gives us all the more reason to speak of an entirely new meaning. It is still the same earth on which we live and work, love and hate, are born and die, but this gives us all the more reason to speak of a new and re-created earth. Humanity and humanizing are seen to be most intensely involved in God's dealings with man. They are the form of the Word which God speaks to man, and are, at the same time, borne entirely by the power of his Spirit, who works all in all (cf. 1 Cor. 12, 6), both to will and to do (cf. Phil. 2, 13).

What is true of the non-human world with reference to human intersubjectivity—that it exists, and must exist, entirely to subserve intersubjectivity—is true of it in a very much more fundamental way with reference to the divine community which is being realized in this human intersubjectivity. Therefore, we must not mistake the sense of certain expressions and tendencies in scripture and in Christian tradition, where these stress that "beatitude" does not consist in wealth, goods, status symbols, position, acclaim, publicity, image, power or physical health. In fact, in this respect, neither does "beatitude" consist in interpersonal humanity, nor does the pessimism of all the things named constitute the meaning of autonomous human life. Nonetheless, just as it all can and must subserve human intersubjectivity, so, in the same function, it must share in the revelation which brings God's presence in the whole human world to light and, in this way, become part of the new earth.

"Beatitude" and *"ultimus finis"* have so far been understood in terms of the fundamental human and divine meaning of man's existence *within this world*. This, perhaps, needs a little explanation. There are indications that words like "beatitude," "perfect happiness," "beatific vision," and other expressions arising in this context such as *"videre Deum," "frui Deo," "contemplatio divina,"* and so on are interpreted as having reference to the definitive state of man after death, and therefore not to the existence of man as *viator* within this world. There is a sense in which this interpretation is right, and another in which it is quite wrong.

The question about the human and divine meaning of life—whence the carefully intended formal significance of *"De ultimo fine et de beatitudine"*—is placed at the beginning of Christian ethics as the fundamental point of departure for a discussion of man's ethos within this world. To take the formulas and expressions that arise in this context, usually derived from biblical sources, and to understand them as referring exclusively to the definitive state of man after death implies that there has been a fundamental misunderstanding of the initial discussion in Christian ethics. For it is with good reason that this discussion is placed at the beginning of the ethics, and not at the end of the whole of theology, under the title *"De rebus ultimis"* ("On the Last Things").

On the one hand, it is established that *"beatitudo," "videre Deum," "frui Deo,"* and similar expressions reveal the divine significance of the intersubjective status of man within the world, so that any other interpretation would lead not merely to the misunderstanding of a portion of theology, but—and this is far more serious—to threatening and distorting the whole content of the Christian message of salvation. On the other hand, it must be admitted that this misunderstanding emphasizes strongly one of the most essential and indispensable aspects of the Christian proclamation. For it is indeed a fact that scripture, the Church's preaching, and theology all very often place more stress on the continuity of the preliminary and definitive phases than on the distinction between what is preliminary and what is definitive. We find an example of this in the preface from the liturgy for the dead: "Life is changed; it is not taken away"; life (see St. John's Gospel) has already begun and is continued through death and beyond.

In other words, preaching and theology sometimes express these thoughts so concisely and realistically that some, in their hesitation and timidity, tend to think that what is revealed must refer to what comes later. Even in doing this, however, they underline another most real aspect: the difference between the preliminary and the definitive which, with right reason—as in all continuity—may not be lost sight of, no matter how much inclination there may be to do so.

There is a difference between *beatitudo* and *ultimus finis,* but it does not exclude the fact that they do, in fact, coincide. We see this in the fact that God's human presence in our world turns human intersubjectivity into community with God—that is, human intersubjectivity is turned into *beatitudo* and into *videre Deum.*[8] The fact that *beatitudo* and *ultimus finis* are different, but at the same time do coincide, will probably help to take away the veil—even if it is different from what is normally meant in this context—which lies over the rather well-known *desiderium naturale ad videndum Deum* (the natural desire to see God).

Many attempts to interpret this desire enter an uneven struggle with the gratuity of the vision of God and with the censured notion of the *"exigence divine."*[9] The most widespread interpretation tries to suggest that it is indeed a natural desire in man, often understood as a kind of desire for infinity, arising from the perception of man's limitedness or from the limitedness of being, but which can be fulfilled and satisfied only by God in perfect gratuity.

To me, this appears to skirt the difficulty by verbiage. It may have some practical value as far as the Church is concerned, but theologically it is of no value. The only sense of the "natural desire to see God," as far as I can see, must be this: in terms of the inmost meaning of man's situation within this world, as brought to light by the proclamation of salvation, every single human and natural desire rightly can and must be understood as a desire for God. In other words, the desire is not one for a kind of infinity that would be sought elsewhere than in being a creature, but a desire for what is actually the deepest—and divinely revealed—sense and content of each and every limited, even if unselfish, desire of man. Misunderstanding of the term *"videre Deum"* and the influence it

[8] Jn. 14, 9: "He who has seen me has seen the Father"; 15, 23: "He who hates me hates my Father also." Cf. Jn. 1, 18; 1 Jn. 4, 12; etc.

[9] Cf. H. de Lubac, *Surnaturel* (Paris, 1946), p. 489; Pius XII, Encyclical *Humani generis,* August 12, 1950.

has had, which we have already mentioned, explains the further complications in which the "natural desire" has become entangled. It is evident that de Lubac, in his two latest books,[10] saw no chance of unraveling the knot.

To conclude this chapter, only one thing more needs to be stressed, and it is this. It is to be hoped that it is now clear enough that a Christian ethic in no way makes an autonomous, human ethic superfluous, but, on the contrary, presupposes it. If an ego-centric human ethic, based on the ideal of happiness, forms the foundation, the only result of revealing the divine meaning of life will be to place God in function of the ego. It is only when one proceeds from the basis of intersubjectivity as the meaning and the task of life that something more will come to light about the selfless generosity of God's love, the love that makes *beatitudo* of our intersubjectivity.

[10] *Augustinisme et théologie moderne* and *Le mystère du Surnaturel* (Paris, 1965).

2. Human Action

This second chapter will be concerned with human action, dividing the discussion into four sections, with a gradual progression in the first three from a more abstract to a more concrete and complete reflection on man's action. The sections will deal successively with (1) freedom, (2) intersubjectivity, (3) good and evil, and (4) what is known as "the passions." This final section is intended to embody the discussion, so to speak, in the reality of flesh and blood.

I

Freedom

One conviction that, more basically than any other, forms part of the Catholic tradition of moral theology is that human action is free. The opposition of the Reform doctrine of (non)freedom served to strengthen the Catholic conviction. Moreover, this same tradition has always attempted to keep the reality in view by taking into account any possible diminution or limitation of freedom. Nevertheless, a conflict arose when modern psychology and philosophy began in their own way to examine the limiting of freedom. In so doing, they used terms such as unfree, determinism, and so on, which to the moral theologian seemed to be a little too much of a good thing. Thus we have good reason to bring some care and circumspection to our approach to the freedom which seems obviously to be inherent in human action.

Is human action genuinely free, or is it determined? Is it a matter of "freedom for," but not "freedom of" action? Is human freedom actually, in itself, a sinful absence of freedom from which

God alone can free a man and make him genuinely free? Is freedom the kind of reality that can grow, or is it the kind of thing that either is totally or is not at all? Is freedom the same thing as intersubjectivity which has succeeded, and is wrong action, then, actually not free?

If we want to find a way out of this maze and give an answer to these questions, two things seem to be important: (1) we must be careful to keep the actual reality in view, for it is not always described or referred to in the same terms; (2) we must give particular attention to the tradition of moral theology, if only to keep open, in that direction too, the possibility of understanding and being understood, and also because there are certain traditional problems that ought not to be neglected.

From the very beginning it should be made clear that in the first instance we are undertaking a human, ethical reflection—human in the strict sense, so that questions about sinful unfreedom and the like simply do not concern us yet. For sin, in one form or another, can enter the picture only in the context of the Christian proclamation.

Let us state a first question, one of essential importance. Should we not recognize that, in speaking of freedom, we are indicating only a particular and limited aspect of that total reality which is human action, especially since we are speaking of freedom only in the line of the ethical tradition in moral theology, although this is also in the line of another and yet older tradition? To be more precise: all human action is surely intersubjective (to be treated in the following section) and good or evil (the subject of our third section), but surely there is nothing to prevent us from giving an especially close look at the freedom of human action—its spontaneity, its personal character, the fact that it proceeds from man himself—and so distinguishing freedom within the totality of that action as one particular aspect of action, in the same way in which we speak of a man as an individual and a person, although we know very well that man is essentially intersubjective and forms part of a certain community. To the eye, a man appears to be a separate, individual person; nevertheless, he is the child of particular parents; he comes from a certain milieu; he has these brothers and sisters; he has received this education; he is married to this particular wife; he has these particular children; he bears these particular responsibilities. In a similar way, human action, in spite of all sorts of real and concrete aspects of relationship to particular persons in a particular situation, has also the aspect of proceeding from the individual concerned *as an individual.* Now this is what is meant, in a long ethical and moral theological tradition, when human action is called free.

There are important consequences to the fact just noted, consequences most closely connected to the anthropological presuppositions of which we spoke in our second chapter of Part I. For if one thinks in terms of the world of Greek culture in the time of Plato and Aristotle, although also in terms of the Graeco-Roman culture of Paul's time (cf. 1 Cor. 8—10; Gal. 5, etc.), then it is easy to understand freedom as a synonym of substance and of independence in a social and economic sense, and also in a human, emotional sense. All such understandings would be in perfect correspondence with the view of man current in those contexts, and with the human ideal, for example, of *apatheia,* of stoic imperturbability and calm. It is difficult to relate a freedom of this sort to the intersubjectivity which we have identified as essential to man; actually it is an entirely different kind of thing. If one thinks in terms of the supernatural, or if one thinks "biblically," then it is easy to understand freedom as opposite to the slavery of sin; but whether any human reality is in this way revealed in its salvational or non-salvational aspect remains an open question.

However, if we take into account all these different connotations which the word "freedom" can have—and does have in particular contexts and in the minds of certain authors—and then say, arguing from an anthropology of intersubjectivity, that human action is free, then our intention is to express and to indicate something that, at least in part, *agrees,* even in its essential, with the use of the word that has its roots in a different anthropology, although, at the same time, our intention is to say something that differs in other respects. In other words, when we argue from a different anthropological standpoint, we ought to be aware that the meaning of our term is altered to some extent, and that we are, to some extent, giving it a new interpretation. There is, of course, no objection to this in principle. Moreover, it is the kind of thing that is happening almost every day. Nevertheless, it is extremely important to be aware of one's own method and procedure and of the fact that one must interpret anew—provided that he is trying to understand the reality and make it intelligible to others, and not merely recording how others have attempted to understand it.

In agreement with the basic tendency of Aristotelian ethics and of the Catholic tradition of moral theology, we have distinguished freedom, within the totality of human action, precisely as the element of its proceeding from the person as such, its spontaneity, or in whatever other way it can be further described.

This is an abstraction, which is to say that, although it is right and correct, it is only part of the truth and that it does *not* express other aspects of human action. However, it also means that one cannot and may not see anything in these other aspects which is in opposition to *this* kind of freedom. The other aspects of human action do not in any way militate against freedom as understood *in this way*.

We can state this more concretely. If we say that human action, as intersubjective, is always bound in a certain relationship to others and to the community, we are not saying anything against this freedom itself. This bond, this relation to others and to the community, is admitted and, in speaking of freedom, it is in no sense denied. We are fully aware that man is essentially intersubjective, and that his action is therefore bound to relate to others, but we are simply not speaking about this fact. When we use the word "free," all we want to do is express this one particular part of the truth, namely, that man's actions proceed from man himself. That he is, in a certain sense, determined by milieu, education, and so on, is quite another facet of intersubjectivity, but, once again, it is not in opposition to freedom, for freedom means precisely that man's actions proceed from himself; it does not refer to whatever other influences may play a role in this.

We are stating the physiological fact of autoactivity, without ignoring the material process of interaction, but above all we are stating the fact of autoactivity as a human and personal reality, again without ignoring on this level the process of interaction in which not only our present environment is involved, but also our wholly individual background and history. For there is no point in speaking of freedom unless it be within the whole of this concrete context.

Now if, remaining within this total context of intersubjectivity, we distinguish in our action the aspect of personal spontaneity or freedom, we must also gradually have become aware of all that is not expressed when we point to this part of the truth. We must have become aware that we owe this freedom entirely to others; that the influence of others, or of force, fear, passion, or lack of attention does not actually detract from this freedom; that this freedom is, above all, a tendency and a charge given to us, a developing and growing reality whose present imperfections we know very well. We know, too, that this alone and this essentially is man, man's situation and man's lot; we perceive that we can underestimate and also idealize another's freedom and that this underestimating and idealizing, in their turn, reveal things about our own freedom and its true condition.

Freedom, then, is a very much more modest reality than that sometimes described in high-sounding and lyrical idealizations; it is more modest, too, than self-assured ethical or moral theological expositions and polemic writings, proceeding strictly on principle, would seem to want us to think. When one does think about it, one begins to think also of good and evil in much more temperate terms. As a matter of fact, one's view of man, in general and in all its detail, becomes more refined and more benign, considerably less blunt and inflexible, considerably less independent and self-sufficient than that view of man and his freedom propagated as a result of a wholly superfluous overestimation of himself.

We have, in passing, already noted certain consequences following upon freedom as understood in this way. But these require closer attention, especially since they play a fairly important role in the Catholic tradition of moral theology.

Freedom—that is, the personal, spontaneous character of human action—is a reality, however modest this reality may be. Or, rather, it is a real aspect within the totality of human action. Now the fact that it is only a single aspect within this totality leads immediately to the possible conclusion that there can be other aspects to that same action, as we earlier indicated, which in no way detract from the aspect of the action's being free, even if at first sight we tend to suppose otherwise.

Here we become concerned with certain points which are classically raised in this context: force, fear, passion, and ignorance. Surprisingly enough, we must note that the average manual of morals accepts these more easily and unhesitatingly as diminutions of freedom, and therefore as possible exclusions of guilt, than, for instance, Aristotle in his *Ethics* or Thomas in the *Summa*. Force, fear, and passion, at least in this context, belong more or less together; to a certain extent, ignorance forms a category of its own.

In the first instance, Aristotle and Thomas, and many others with them, see in force, fear, and passion no diminution of freedom. Force (*vis*) is not to be understood as only physical overpowering. In a certain sense it applies as well to all the inevitabilities of life up to and including the inevitability of "the other" who, uninvited, intrudes his presence into my way of life, or who, although invited, proves to be something he was not asked to be. Fear is generally understood as dread (*Angst*), although there are clear indications that Thomas, for example, mentions the factor of fear apart from and alongside of the factor of passion (which should include fear in the sense of dread), only because in this way is he able to indicate an entirely special aspect of human action. This is the aspect of its being faced with a choice and placed in a dilemma which demands a decision—often a quick decision—in one particular direction, and which therefore means that other possibilities must be sacrificed and lost. The classic example of the captain of the ship who, in the face of a violent storm, must throw the cargo overboard to save the ship and the crew can be replaced by other situations in which man is faced with a dilemma. Some examples: either the mother or the child (although this case, too, has become practically archaic); either the atom bomb or a probably interminable war (as in 1945); political and military decisions on which men's lives and incalculable risks hang in the balance; and so on. Alongside of inevitability and dilemma, there is, finally, the physiological phenomenon of passion.

Now, in general, the factors gathered under these categories are not taken (by Aristotle, Thomas, and others) as detracting from freedom or diminishing it; they are, rather, regarded more as a challenge to the exercise of freedom. As far as passion is concerned, it is seen, in a certain sense, even as an enlargement, an increase of freedom (except in cases of pathological factors which cannot be integrated).

The reason for this view must be sought in the explanation of freedom that is coupled with it. A certain limiting and determination of human possibilities lies in the very nature of human intersubjectivity. The possibilities open to man, or man's freedom, consist precisely in the fact that he himself is able spontaneously to choose a position regarding the human situation which he faces. He can adapt himself to the inevitable, but he can also rebel against it. He can bow fatalistically under difficulties, but he can also try gradually to overcome them. He can avoid certain of his fellowmen, but he can also make efforts to reach an understanding with them. In other words, freedom does not consist in being able to design a world for oneself; rather, it consists in being able to take up one's position for or against a world which, in a certain sense, is already designed and forces itself upon one.

In this connection, ignorance plays a different role. As an expression of this role, our word "ignorance," as also the Latin "*ignorantia*," is not very satisfying or exact. However, if "ignorance" is taken in the sense of "not knowing," then it can serve to describe one particular and limited aspect of its role, which one can find fairly well illustrated in an expression such as, "He did not know; he cannot be blamed."

In the present context, there is not much point in mustering all the details and distinctions of "antecedent," "concomitant," and "subsequent" ignorance, of "vincible" or "invincible" ignorance, and so on. Two main considerations are important for an understanding of freedom.

1. Simple ignorance, in the fundamental sense of "not knowing," means that that knowledge which is one of the necessary conditions for an action really to proceed from man himself—that is, for an action to be free—is lacking.

2. Very often, what is called "ignorance" is, in fact, lack of attention, nonchalance, carelessness, lack of a sense of responsibility, and the like. These do not in any way detract from freedom. On the contrary, they are elements of freedom, and form part of it.

We have sketched the main outlines of freedom. Against this background it will already have become clear what sort of thing we find in the manuals of moral theology: a view of freedom that we might well call optimistic; yet, rather surprisingly, alongside this view there runs something that is foreign to, and even in conflict with, the great ethical tradition: a certain bland and unquestioned acceptance of the notion that factors such as force, fear, passion, and ignorance detract from and diminish freedom—hence their classification as "*hostes voluntarii*" ("enemies of freedom").

Most noticeable in this so-called practical moral approach is that, as always, it does in fact come to conclusions which seem to be practically the same as those in the great traditional line of ethics and moral theology. Where, on the level of pure principle, something is added to freedom, it is nibbled away again on the level of the concrete instance under pressure of the actual circumstances (not to say, of the situation). Accordingly, just about as much finally remains as in the theoretical ethic that, on the level of principle, expresses its view on human freedom with great reserve, and that, on the level of the concrete situation, is therefore not forced to go back on these views. Unlike this, the practical type of morals is always only too ready to be benign, to look for grounds of exoneration, and especially to call into doubt whether there actually was any "full knowledge" and "full consent." We can therefore frame the question—too wide in its implications to be investigated here, but sufficiently relevant to be stated—whether the modern concept of freedom does not conceal, in all the vagueness that adheres to it, an individualism and an idealism which will have to be unmasked once again in casuistry and jurisprudence where the factors which have limited and diminished freedom are unraveled and brought forward. It may be helpful to realize that the French Revolution of 1789 and the anti-colonial revolution after 1945 did indeed alter situations, but did not change man. Stated in a more general way, the word "freedom" carries so many historical connotations that here, too, one is tempted to identify ethos and ethic.

An extremely realistic, but at the same time modest, view of freedom does not see in freedom one or another kind of separate and distinct quantity, but simply one aspect among the many aspects of the total reality of man in action. It sees freedom as the issuance of action from the person himself and, in this sense, as the spontaneous character of human action, without opposing this in any way to the dependence, the situation, the limitations, the inevitabilities, and the dilemmas with which man, in his life with his fellowmen, is confronted.

II

Intersubjectivity

In spite of the fact that we do not yet touch on the question of good and evil, we shall, under this heading, deal with the most essential matter in the whole of morals. Here we find ourselves, as it were, in an intermediate phase. On the one hand, we have treated the question of freedom as the aspect of action which specifies it as spontaneous and personal—in a certain sense, the most abstract question before us; on the other hand, we must still arrive at the concrete and total human action which must be either good or evil.

How do custom and tradition speak of human action? That is, how is human action treated within Catholic moral theology and outside of it?

In the manuals of moral theology we find the treatise on freedom followed immediately by one on the morality of human action. By "morality" the manuals do not mean that aspect of human action of which we now want to speak; rather, they speak of an action as good or bad. In our development, this point will not be raised until we come to the next section. We do find in the manuals, nevertheless, a very brief word on intersubjectivity and, in this sense, on the specifically human character of action; it is either included in the chapter on freedom or it opens the chapter on morality. The distinction is made, however summarily, between *esse physicum* and *esse morale,*[11] between *actus eliciti* and *actus imperati,* and sometimes also between *actus in finem* and *in ea quae sunt ad finem;* again, this procedure may occur in connection with the analysis of the human act into eleven or twelve different acts. In addition, one probably comes across the distinction between the object of an action and its circumstances, although, as a rule, this appears under the heading *"De fontibus moralitatis."*

In other words, the point in which we are now interested seems to have slipped down between the pier and the boat, at least as far as the manuals are concerned. To put it in another way: these books do not take the factor of intersubjectivity as a point to be treated apart and for its own value, but almost always in combination with, and as a uniform part of, the question of good and evil.

What we have noticed in these books, we can also notice in the conventional framework of our conversation on this topic. When people speak about means and ends, they are actually touching on the matter that concerns us now, although the terms "means" and "end" are used, as a rule, in conjunction with the qualification

[11] D. M. Prümmer, *Manuale Theologiae Moralis* I (Barcelona, 1958), pp. 67–68; B. H. Merkelbach, *Summa Theologiae Moralis* I, (Paris, 1938), p. 106.

"good" or "bad." For instance, it is said that we may not use a bad means in order to achieve a good end. When someone's theories are said to imply that the end justifies the means—although the charge is often more political than scientific—that person is suspected of being Machiavellian or of being tainted with "Jesuit" morality.[12]

Concrete problems of this kind are familiar enough. Birth regulation is a good end, but sterilization (by means of mechanical or chemical contrivances, whether old-fashioned or new) is a bad means. To provide someone with a necessary organ by transplantation is a good end, but to maim the donor by depriving him of that organ is a bad means. It is good to have children, but artificial insemination is not a good way to go about it. It is good to heal mental disturbances, but you may employ no prohibited means. All this we find not only in the manuals of morals, but also in the way we normally speak.

In other words, inventory in this field reveals that no clear separation exists between the aspect of intersubjectivity and the question of good and evil. The question about the proper nature and specific character of human action is indeed, in a certain sense, an abstract question, just as the question about freedom is abstract. For here, too, we are still concerned with one aspect of human action, an aspect that is common to both good and evil action, just as freedom is common to both. We have not yet come to consider the concrete action itself. The concrete act itself is *either* good *or* bad, if it is a genuine human act and to the extent that it is a genuine human act. For this reason the question of freedom and the question of intersubjectivity are of essential importance. As a matter of fact, good and evil refer to the success or failure of intersubjectivity, and for this reason there cannot be any question of good or evil unless there is first a question of intersubjectivity; furthermore, we may speak of good and evil only to the extent that we speak of intersubjectivity. Thus the problem—we might even say the disease—with the usual approach to morals is that it begins too soon to maneuver among categories of good and evil, and so its conclusions are premature. For they can be no more than a shot in the dark unless the factor of intersubjectivity has

[12] Cf. E. Weil, "La morale de l'individu et la politique," in *Tijdschrift voor Philosofie* 27 (1965), pp. 476–490.

been investigated. It will be clearer through our subsequent discussion that no implication is here intended to the effect that there is nothing of value in this approach to morals.

One of the few matters sometimes presented in these manuals of morals as a preliminary to the question of good and evil is the distinction between the object and the circumstances of an act. For example, the object of the act of fasting is simple abstention from food, while a circumstance of the act is that it is to the honor of God; or, the object of the act of killing is the termination of a life, while a circumstance of this act is that it took place in a church, thus making it a sacrilege. The terms of this distinction are also known as *"substantia actus"* and *"circumstantiae."* This distinction usually appears in connection with the question of good and evil, is often confused with it, and only too seldom is considered deserving of separate treatment. Even so, it does offer us an important avenue of approach to the point that concerns us now, especially since we are trying to take into account the role which this distinction between the object, or substance, of the act and its circumstances has already played in our current concern.

The catechism states that the number and the circumstances of our sins must be told in confession, a ruling, as is fairly well known, that dates back to the Council of Trent. The 14th session of that Council produced a decree on confession which states explicitly that one is bound to confess circumstances "which alter the nature of the sin."[13] Equally relevant, however, is the historical development which formed the background to this decree and in which reflection on human action played a central part. From the 7th to the 12th centuries, the confessional and penitential practice of a large section of the Western Church was officially regulated by the *"libri poenitentiales,"* the so-called Books of Penance.[14] As we know, these Books of Penance, a kind of little handbook for priests, contained a catalogue of the most prevalent sins, together with precisely determined penances to be imposed. Here originated the expression "tariff of penances." For more reasons than one, these books were compared to a code of penal law—whether

[13] ". . . *quae speciem peccati mutant*": *Conc. Trid.* VII (Freiburg, 1961), p. 348 (cf. Denz. 1681, 1707); also cf. *Tijdschrift voor Theologie* 4 (1964), p. 161, n. 28.

[14] For a short bibliography, see *Tijdschrift voor Theologie* 4 (1964), pp. 262–263.

ecclesiastical, patterned after the civil code, or without any clear distinction as to the civil or ecclesiastical nature of the code.

In practice, however, the application of this "code of penal law" was much less severe than the texts themselves might lead one to suppose. Bishops and synods on the Continent repeatedly protested against the use of and the mentality behind these books, most of which were imported from Ireland and the British Isles. But apart from this, possible subsidiary penances were appended to the tariff penances in the books themselves, and these, known as "redemptions" or "commutations," provided a diminution of the penance. The flexibility of the system in practice was more strongly accentuated when, from the beginning of the 11th century, the introduction of so-called *indulgentiae* not only brought about a more benign application of the penitential tariff, but, above all, placed—or replaced—the chief stress on the "conversion" of the sinner rather than on the external penance. What were these "indulgences"? In principle, the indulgence could apply only where a conversion or change of inward attitude was already evident; that is, it could apply only to those already "in the state of grace." It replaced the extremely severe penance called for by the tariff (fasts of many year's duration for just one particular sin were quite normal) with the imposition of a much lighter task, usually of a social or charitable nature (a contribution toward the cost of a bridge, a church, and the like).

By the 12th century—certainly by the 13th century—this development had led to the total disappearance of the tariff penances and to their replacement by *poenitentiae arbitriae*. These were not arbitrary penances, but penances imposed, in view of the situation of the penitent, according to the judgment (*arbitrium*) of the confessor. (By this time, indulgences had already, in fact, outlasted their useful purpose. Their further development is to be explained partly by the "traditionalism" of both hierarchy and people, partly by the fact that those engaged in "speculative" theology no longer knew the historical background of indulgences. In this matter, Thomas—once again—is an exception.)

The theological reflection of the time on this whole development in the Church manifests elements that relate precisely to the question that concerns us now. Theologians protested and reacted against the one-sided and external view of sin and of penance found in the Books of Penance. Abelard goes so far as to state that it is not the external act, but only the *voluntas* or *intentio* that has any significance. Against the background of certain of Augustine's writings, on which Abelard claimed to base himself, his view implies a great deal more than our word "intention" is capable of suggesting. But to the 12th-century ear of William of St. Thierry and St. Bernard of Clairvaux, it was such a departure from sound doctrine that it helped toward Abelard's condemnation by the Council of Sens in 1140 (or 1141).

Abelard, however, had clearly touched a sore spot. The problem he had raised immediately took the center of the stage in theological discussion and remained there until it came to a provisional conclusion and achieved a measure of official recognition in *"Omnis utriusque sexus fidelis,"* the famous 21st canon of the Fourth Lateran Council in 1215. This canon prescribes yearly communion and confession for every one of the faithful, but at the same time (with clear allusion to the parable of the Good Samaritan) it describes the task of the confessor as that of the *"peritus medicus."* He will not mindlessly impose uniform penances for outwardly uniform behavior, but he will take careful account of the actual "circumstances of the sin and of the sinner" and, in so doing, will pour oil and wine on his wounds to help toward his healing in the manner that is fitting.

This definitive approval of the *poenitentiae arbitriae* amounts essentially to an explicit recognition—although, in one sense, an opposition to certain tendencies of the Books of Penance—of the truth of the old Roman proverb: *"Duo dum faciunt idem, non est idem"* ("Although two do the same thing, it is not the same thing"). It is not of any great account that this conviction continued to be referred to in terms of act (or *substantia actus*) and circumstances, terms which the theology of the time had borrowed from classical rhetoric. What is really relevant in this famous conciliar canon and, at the same time, what manifests a development that is both human and ecclesiastical is the fact that the unique character of human action was recognized, and that the practical instruction of priests was adjusted accordingly. It is admitted, even if the terms of the admission leave much to be desired, that human acts which are entirely different, one from another, may possibly be externally and materially the same.

Against the background of the penitential practice and the Books of Penance in the early Middle Ages, the pronouncement of the

Fourth Lateran Council is a real milestone. When Trent, 350 years later, spoke of confessing the circumstances that altered the nature of the sin, it was only repeating what the Lateran Council had already said, and the manuals of theology and the catechism have continued this down to our own day.

To focus more sharply the importance and the ramifications of the ethical problem which arose here in an explicitly theological and ecclesial context, it helps to know something of its development from the Fourth Lateran Council to Thomas. During the whole of the 13th century, theology waged a constant struggle to penetrate the problem more deeply and not to be content with so provisional an answer as the distinction between object and circumstance. This distinction, however, did represent an enormous advance when compared with conclusions based merely on what is material and external in an act and which were, not unjustly, considered to be characteristic of all that was bad in the regime of the Books of Penance. However, we shall not relate the full history of this question, but shall keep simply to those elements which matter to us now.

This history provides us with the categories of object and circumstance, or of act and circumstance (for instance, stealing, stealing in dire need, stealing without need; killing, killing in self-defense, and so on). These categories provide us with one set of tools for analyzing the human act. The categories of means and end provide another set of tools. History shows that both sets have an imposing record of useful service, and the conventions of our present language show that means and end are at least as current as act and circumstance. The reality itself, however, is much more important than categories and the tools they provide, and when we do gain insight into the reality itself, these categories and other ways of approach will themselves become more intelligible.

In our chapter on man (the second chapter of Part I), we spoke of corporeity and intersubjectivity. Man can be referred to and qualified both as corporeal and as intersubjective. The two qualifications refer directly to precisely one and the same reality, but each in its own way and each from its own point of view (as, for example, both Roncalli and Pope refer to one person, and both Johnson and President refer to one person, the former in each case referring to him as an individual, the latter as the holder of office). Corporeity qualifies man under all the aspects in which he coincides with and forms part of the non-human world; intersubjectivity points him out in his human uniqueness, as, in a measure, the word "freedom" has already done.

Now, if something is true of man, it must be equally true of his action. For this action is indeed nothing else than the concrete manner in which a man exists and in which he is corporeal and intersubjective. We can apply the distinction between corporeity and intersubjectivity to man's action as well as to man himself, and it follows immediately that this is the most fundamental thing to be said about human action. It is without doubt a reality which is wholly corporeal, yet we encounter its uniquely human character only when we look upon it as intersubjective.

Some examples may help to clarify this. In the first place, recall the case of walking as muscular movement, biochemical reaction, consumption of energy, and so on, as distinct from going for a stroll, going to work, accompanying someone, swaggering, and so on; or the case of moving lips as muscular movement and the like, as distinct from speaking, kissing, tasting, or whatever. Other examples, however, more quickly and more clearly bring out the implications of the distinction between corporeity and intersubjectivity. The physical, bodily reality of killing can be, as an intersubjective reality, murder, waging war, administering a death sentence, suppressing an insurrection, self-defense, and so on. Taking something from another can, intersubjectively, be stealing, borrowing, satisfying a dire need, repossessing, and so on. Intersubjectively, failing to tell a truth or telling an untruth can be deception, joking, keeping silent about things that do not concern another, and so on. Masturbation, intersubjectively, can be the satisfaction of a sexual urge, experimentation, or, in the case of a man, the provision of semen for scientific research, and so on. Removing a fetus from the womb before it is viable can be abortion or murder, the removal of the effects of rape, saving the life of the mother, and so on. Intersubjectively, genital sexual activity between man and wife, whether "completed" or not, whether with the use of some kind of contraceptive or not, can be love, egoism, self-satisfaction or aggression. To use a knife on someone can be, intersubjectively, having a fight with him or the surgical removal of an organ for transplantation.

Practically all the examples given, it will be noticed, have a history, short or long, perhaps stormy (in some cases very stormy), and some are still the subject of controversy. There are many reasons for this. Although one must not be misled into a far too simple view of these problems, there is nothing to prevent us from establishing, simply and from the very beginning, the fact that numberless problems and interminable polemics arise, and have arisen, through neglect and disregard of the entirely unique and special character of human action. So often the material, bodily reality of the action is unthinkingly and wholly identified with one single form of intersubjectivity to the total exclusion of all others (tantamount, for instance, to saying that movement of the legs can be only a prim gait and nothing else).

This narrowness of outlook results partly from the fact that the qualification "good" or "bad" is taken to apply directly and without further ado to the corporeal act as such, whereas, in fact and in the nature of the case, good or bad can be involved only in the *human* act—the act as intersubjective. Intersubjectivity as a factor is simply ignored. The most that can be achieved by way of refinement consists in the recognition of special circumstances. However, these circumstances are thought to make no difference to the essential act (*substantia actus*: the object of the act) as such (for example, killing or killing in punishment; stealing or stealing in dire need).

This explains why Catholic moral manuals are, by their very nature, conservative—ultraconservative—not only in a good sense, but, unfortunately, also in a bad one. That the same material, bodily act may possibly have a *different* intersubjective significance is something that, in principle, lies outside of its field of vision. Abortion is murder; the possibility that a termination of pregnancy could be medically indicated is something that simply cannot be taken into account. A surgical intervention is either therapeutic or it is mutilation; the fact that transplants involve something entirely different cannot even be considered. For, indeed, the act itself, the *"substantia actus,"* cannot be altered.

The Lateran canon and its accompanying theology brought about a favorable change of direction at the beginning of the 13th century. For one thing, however, this remained too superficial; for another, the terms used (act and circumstance) gradually became so simplified and inflexible that the situation appeared once more to be practically as it was before the Lateran Council, in the halcyon days of the Books of Penance with their interpretation of the mere external act in one exclusive sense.

We may ask whether the use of the categories of "means" and "end" would not have provided better opportunity for understanding intersubjectivity. In clear fact, however, this was not the case. In the chapter on man, mention was made of means and end as formally and not materially distinct, means being related to end in the same way as corporeity to intersubjectivity. *Only if* means and end are understood in this way can they serve as an excellent formulation of the reality of human action that, at one and the same time, is both corporeal and intersubjective. For example, termination of pregnancy could be called "means," and intersubjectivity would be indicated by "end," whether it be murder, removal of the effects of rape, or saving the life of the mother. Sexual intercourse of one type or another would be seen as means, the intersubjective end being egoism, or love, or self-satisfaction, and so on. The extreme importance of finding a clear way of expressing these fine distinctions will appear in the next section, the section on good and evil.

In practical morals or ethics dealing with concrete situations, as well as in the conventions of our daily speech, the categories of means and end are used and understood in a more material sense. Their case is rather like that of the categories of act and circumstance. Means, or material act, are identified with intersubjectivity. In other words, any view of a *different* intersubjectivity or of a *different* human significance in the act is, so to speak, excluded *a priori*.

There is yet another set of categories, that of act and intention (or purpose). What we have said about the category of means and end, we can say here, too. Normally, we look upon the act as something unchangeable, something established, and upon the intention or purpose of the act as something that exists apart from it, as something that can freely be added to the act from without, at the same time actually doing nothing to change the act itself. Accurately speaking, however, we should have to say that act and intention coincide materially, that they are, in reality, the same thing, just as corporeity and intersubjectivity, means and end are the same thing, although formally distinct. In other words, act refers to the whole action as a physiological reality, while intention refers to the same action, but precisely as human and intersubjective. Walking, for example, is corporeity, or means, or act; mincing primly is intersubjective, or end, or intention. And so on.

On this account, one can also apply the same method of approach to the category of act and circumstance—Thomas pointed out this fact[15]—particularly if it is remembered that *quid* and *cur,* which are, respectively, act and intention, are taken to be the chief "circumstances" of an act.

When Thomas distinguishes between *"secundum speciem naturae"* and *"secundum speciem moris,"*[16] the same distinction between corporeity and intersubjectivity is intended, but we can no longer—or just barely—find this distinction in authors like Prümmer and Merkelbach. Although they do make a distinction between *esse physicum* and *esse morale,* they usually understand these terms in quite another sense.

Finally, the act having a double effect provides categories which, like those we have seen, were meant to be used in treating of the same problem. Alongside the proper (good and willed) effect of an action is a second (evil and not willed or not directly willed) effect which demands a justifying reason. A recent example of this is found in the use of progesterone and similar preparations with both a therapeutic effect and a secondary sterilizing effect. This principle has probably helped most toward meeting the problem of means and end, act and effect, and so on, because it extrapolates, as it were, means and act to the second effect. This helps to bring out the corporeal character of the action (act) and to distinguish it from the intersubjective character (effect), although it does not do so very clearly. The outcome is obscured, however, because a twofold intersubjectivity is ascribed to the act (two effects), and then one aspect of it is immediately canceled out ("indirectly willed" for "sufficient reason").[17]

Whatever words we choose to use and however we interpret them, the most important thing is obviously the human reality it-

[15] *Summa Theologiae* I–II, 7, 4; see *Tijdschrift voor Theologie* 4 (1964), pp. 160–161.

[16] *Summa Theologiae* I–II, 1, 3 ad 3; cf. II–II, 59, 3 ad 3.

[17] For an analysis, see P. Knauer, S.J., "La détermination du bien et du mal moral par le principe du double effet," in *Nouv. Rev. Théol.* 87 (1965), pp. 356–376; for a historical study, see J. Ghoos, L'acte à double effet. Etude de théologie positive," in *Ephem. Théol. Lov.* 27 (1951), pp. 30–52.

self, and this reality is what we need to understand. If we are aiming at a profound perception of human action, we must clearly recognize that the factor of intersubjectivity demands special consideration before we can speak meaningfully about good and evil.

The way things are in principle—and the way we have to consider them fundamentally, if we are to gain any insight into the complex reality—is that what is material in human action is able to be intersubjective in the most diverse and varied of ways. Nevertheless, the reasons why moral theology has so often identified a particular mode of intersubjectivity with the corporeal act (for example, interruption of pregnancy with abortion or murder), without taking other possibilities into account, are not always to be sought in moral theology itself. On the contrary, this moral theology as a casuistry is, in a certain sense, a normal and even a strikingly useful sociological phenomenon—even if one must say at the same time, and once again, that it certainly does not represent any fundamental and scientific ethic.

Actually, what is the case? In principle, it is true that corporeity is able to be intersubjective in any number of diverse and varied ways, but we ought to keep very clearly before our mind the fact (as noted in our introduction) that practically the same holds true of corporeity as a whole, as is very evidently and widely known to hold true of speech, a very limited part of this whole corporeity. Only in virtue of the language which is current in our own milieu, and which we have learned, and learned to use, from our earliest years are we able, in fact, to achieve any intersubjectivity and communication.

We can state this in another way. We are not free to choose arbitrarily the sounds we make, if we value normal communication with our fellowmen. Or we can state it in another way and say that we are not condemned perpetually to seek anew and put to the test ways in which we can turn our corporeity into intersubjectivity. Both these statements of the case are relevant, not only to language as speech, but also to the whole terrain of corporeity and intersubjectivity. For speech, or language, in the strict sense is only one small facet of "language" in the much wider sense, as all that makes our intersubjectivity and communication possible and, at the same time, subjects it to a number of restrictions. All our social institutions, patterns of behavior, table manners, clothes, styles of writing, traffic regulations, and so on, up to and including

Halloween pumpkins and Christmas cake, New Year's Eve, types of children's games, and things like that—all these belong to "language" in the wide sense.

Casuistry, then, with its long list of outward acts, each having but one meaning, is, to a certain extent, a mirror of our social "language." Although this is, in principle, a recognition of the positive value of casuistry, still, flexibility must be included and preserved even on the level of the concrete. Indeed, even our habits of speech are constantly changing: new words are formed and old ones fall into disuse; words may have more than one meaning, and the sense may depend on the stress or the context in which they are used. One can speak with irony or with sarcasm, and so on.

When we turn to a consideration of human action, we ought to take all these aspects of the reality into account. Human action is both corporeal and intersubjective, but the range of possibilities that this fact opens up in principle is always rather powerfully narrowed and determined by the social milieu. It is this last fact that finds overabundant expression in the casuistic approach to morals. But if one loses sight of the relation, in principle, of corporeity to intersubjectivity, of means to end, of act to intention, then one is, in principle, forever bound to one particular "language," without the possibility of amending it in any way, without the possibility of taking over words from another language or forming new ones.

To see that human action is essentially intersubjective is not to have solved any moral problems. One must be on guard against cherishing this illusion. But it is a starting point from which one is able to gain a real and unprejudiced view of human action and of its many possibilities. It is, moreover, a starting point from which one will be able to break through the impasse in which the casuistic approach to morals will inevitably find itself, without having to attack the positive value which casuistry does achieve and without an iconoclastic need to break it down because it is an image and likeness of the current "social language."

III

Good and Evil

Intersubjectivity, as a rule, is understood to be synonymous with a fulfilled openness of one's own humanity to the humanity of others, with love, with communication, and so on. Thus, it is usually understood in the sense of fulfilled intersubjectivity rather than unsuccessful or frustrated intersubjectivity. In spite of this, we have usually used the word in a wider sense, especially in the previous section. The intersubjectivity of which we now speak is not yet qualified as "fulfilled" or "unfulfilled," and it indicates that aspect of human action which is common to love and to hate, to inclination and to aversion, and which is, therefore, common to both good and evil.

The fact that human action is intersubjective means that it necessarily has consequences favorable or detrimental to the mutual relationship of the persons concerned. To state this more directly, intersubjectivity is a form of either communication or the disruption of communication; it is a form of either community or the destruction of community. When we now speak of act and consequences, of act and effect, of means and end, we are, in the first place, not speaking of something that happens *now* and has results, consequences, or effects, or that achieves an end *later;* rather, we are speaking of a particular corporeal action that, precisely as a human act, has immediate implications with respect to the relationship between "subjects."

Now the essential significance of the words "good" and "evil" is simply a qualification of these implications, effects, consequences, or results. In other words, it is simply a qualification of the human content of the act. To call something "good" or "evil" is therefore, in the first instance, a highly pragmatic statement that can be made only after the event, after one has been able to establish the "results" actually produced by the action. In other words, we cannot call an act "good" or "evil" in virtue of the material content of the act, but only after its human significance and content have been established.

This amounts to no more than a concise and summary formulation of something that, as a human reality, presents itself in a much more varied and complex way than any theoretical analysis is able to express in a few words.

Such a formulation gives rise to questions, and the first is probably this: Are we using the terms "good" and "evil" merely as qualifications that can be determined only after the event? Is it not a fact that there are many acts about which we can state, in advance and absolutely, that they are good or bad, or of which we can say that they are good or bad in themselves?

This is undoubtedly the case, not only in our conventional way of speaking, and at least implicitly in civil legislation, but also in ethics, in moral theology, and in scripture. In these contexts killing, stealing, telling lies, and such are called "bad in themselves"; helping the poor—giving alms, as it is classically known—nursing the sick, and the like are called "good in themselves." Yet a third category belongs in this scheme of division: the so-called neutral or indifferent act, such as walking. The classical example here is *"levare festucam de terra"* (picking up a twig).

We may say, then, that this is the classical division: good, bad, and indifferent. Certainly this division has been in general use since the time of the Stoics, and the terms are used in the works of the Greek Fathers and throughout the whole Latin tradition.

The use of these categories did not mean, however, that no problems remained unsolved. Earlier generations of theologians and writers on ethics perceived only too well that murder and capital punishment were not the same; in other words, they saw that killing, although bad in itself, could sometimes, in some way or other, apparently lose some of its badness. Again, helping the poor was not so good in itself that it was impossible to help the poor in a bad way to benefit oneself. Likewise, one could pray or fast merely to be seen by men.

In practice, a solution to these problems was always found, although the theories behind the solutions were sometimes very different. The 13th-century theologians chose perhaps the most circumstantially detailed method in their attempts to find a way out of the difficulty. Their distinctions, divisions, and subdivisions have certainly contributed unpleasant, often detestable undertones to the word "Scholasticism." Among the least extreme of these theologians was William of Auxerre who made the following distinctions: *"bonum in se et secundum se,"* such as loving God, which could never be bad; *"bonum in se, sed non secundum se,"* such as helping the poor, which could sometimes be bad because of a "bad intention," and so on. Corresponding to this, he has, of course, a *"malum in se et secundum se,"* and a *"malum in se, sed non secundum se,"* and so on again. Elsewhere we find circumstances offering an outlet by changing the object, and elsewhere again it was the act with two effects, of which the bad one was not directly intended and did not weigh against the good one.

With the above discussion of intersubjectivity in mind, it should not be difficult to understand the problem just outlined. The threefold division of good, bad, and indifferent, with all its *a priori* qualifications of acts as good and bad in themselves, is on the same plane as the "social language," which we have seen to be the plane on which the concrete ethos and the ordinary conventions of language as well as special ethics or the average moral theology have their function and place. This means that when killing, stealing, and telling lies are said to be bad in themselves, it is implied that, within this community, this meaning is given *a priori* to the material, corporeal act described in these words; it is implied that this community attaches, *a priori,* to these actions a significance that makes them unacceptable as normal behavior, just as in the spoken language certain words have no place in normal conversation.

Casuistry in morals and normal usage in language exist and function on the practical social level, and it is right that they should. We cannot object to this. On the contrary, it is absolutely necessary that we know, at least in general, what must be done to achieve a particular result (in terms of community and communication) or to know what the (human) result of a particular way of acting will be *before* we actually proceed to act. However, the limitations of functioning on this level should also be noted and understood, for here one is concerned with the prevailing forms of behavior and the actual "language" spoken, things that are, practically speaking, of the utmost importance. But on this level one simply does not ask any question about the essential and fundamental significance of these forms, and this can easily give rise to conservatism, the demand for conformity, conflict in encounter with other forms of behavior, and the like.

Investigation and explanation of the fundamental significance of the common forms of behavior is what we mean by anthropology and fundamental ethics. These sciences examine and reveal the significance and content of good and evil. Ethics is not opposed to the "social language"; it is not opposed to the forms of behavior in actual use. On the contrary, it reflects on this "language" and recognizes its necessity, for without it a normal community cannot function and normal communication would be impossible.

Neither is ethics opposed to labeling actions as "good," "bad," or "indifferent." On the contrary, ethics reflects on these qualifications and recognizes that, in reality, they fulfill the functions of recommendation or warning, and that in this way they are a kind of legislation "before the letter." Ethics recognizes reflectively that this is the reason why the terms "good" and "bad" can sometimes have a certain magical undertone. For the community ingrains its customs so deeply and so sharply in its members that people begin to look upon good and evil as inherent in the external act itself; it is better, some may say, to eat no meat on Friday, even though the rule has been done away with. Some even regard the very fact of doing away with the rule as not entirely in order. (Further reflections in this direction, on the super-ego and matters of this kind, are not within our competence.)

Above all, ethics in its reflections recognizes the deepest significance of the terms "good" and "evil," terms which on the level of the "social language" are still only of pragmatic value: this you may do, that you may not. Good and evil in the deepest sense are, respectively, constructive of community or destructive of community, a character which belongs specifically to all concrete human action and which is not simply and absolutely confined to the particular, limited forms of behavior of any one community.

The threefold division of good, bad, and indifferent is an almost infallible indication that one is still moving on the level of the "social language." A closer definition of good and evil is more or less inherent in our first question about the good-bad-indifferent trio. Until now, we have simply noted the fact that people work with the categories of good in itself, bad in itself, and indifferent as applied to external acts.

Practical moral theology does not rest content with these bare qualifications which, in fact, are simply borrowed from the prevalent culture or perhaps even taken from scripture. It also attempts a closer analysis of good and evil, as well as a theoretical investigation of when an act should be called "good" or "evil." In doing this, it is in a sense exceeding its own possibilities, but it is quite ready to do this without a blush for several reasons, one of which is the firm conviction that this is following in the footsteps of St. Thomas. While this is a rather theoretical matter, still, for two reasons, it calls for a moment's attention: on the one hand, because we are concerned here with categories which are still in frequent use; on the other hand, because an unfounded illusion is cherished here, in support of which the good name of Thomas is taken in vain. In a particular context, Thomas may well have been called by his biographer a "dumb ox," but this does not mean that he can be yoked to any old cart with impunity.

The theoretical determination and analysis of good and evil in practical moral theology, of which we have been speaking, contains three elements in particular, out of which the good or the evil of an act is, as it were, constructed. More accurately, a good act is said to contain all three of these elements; a bad act is bad because at least one of the elements is missing or is bad.

These three elements are object, circumstances, and aim. One immediately recognizes the scheme: object–circumstances or act–circumstances that we saw in the previous section on intersubjectivity, in connection with the Fourth Lateran Council of 1215. In this Council the scheme marked an important advance away from an approach based too exclusively on the outward appearance of acts, an approach of which the regime of the Books of Penance had been accused.

Into this scheme the "act–circumstance" factor is now added: the aim or goal or intention. In itself, this is already remarkable, because, in a sense, the goal or intention belongs among the circumstances, or at least belongs in a scheme of the same sort as "act–circumstance," but then with the terms "means–end" or "act–aim (or intention)." Thus there is something puzzling about this scheme of "act–circumstances–aim" or "object–circumstances–aim." It is not really possible to explain this duplication of the last factor which has here become two factors: circumstance and aim.

Writers in practical moral theology claim that this division comes from Thomas.[18] In so doing, they obscure the sense of the first of the four articles in which they claim that these three elements are discussed. In any case, the appeal to Thomas is entirely in vain, as I have shown elsewhere,[19] because the sense of these particular articles is entirely different from and much more fundamental than the sense presented by the superficial interpretation often in general circulation. In that particular place, Thomas is simply showing that

[18] *Summa Theologiae* I–II, 18, 1–4.
[19] *Tijdschrift voor Theologie* 4 (1964), esp. pp. 166–167.

the various distinctions already current in his day (such as "metaphysical" good, object, circumstance, aim) are merely preliminary attempts to solve the problem we were discussing above. Thus, the threefold division, as well as the appeal to Thomas for its authenticity, falls away.

What we have just said answers the well-known question of the so-called indifferent acts. A concrete act, because it is always intersubjective, can never be indifferent; it can be only either good or bad. The category "indifferent" is thus, once again, a category that refers to the level of the "social language" and to the currently ruling forms of behavior. In connection with the question about an action which would be indifferent in itself, Thomas remarks that only an act which is *not a human act* can specifically be called indifferent.[20]

The famous question of means and end, and especially of bad means toward a good end, can be taken care of in this context of good and evil. In the above discussion it will have become clear on what level and in terms of what presuppositions we may speak about a bad means to a good end. In this, once again, we must consider a manner of thinking and of understanding that is entirely dominated by the social pattern actually in force. Within this social pattern, the external act has already received a determined meaning or, at least, is presumed to have received a determined meaning exactly like a particular word in spoken language. Thus it happens that the act is qualified in advance as good, bad, or indifferent.

A fundamentally incorrect way of speaking about means and end is one which does not correspond to human action as corporeity–subjectivity, and the social labels attached to acts lead to the formation of an ethical verdict of rejection or to the setting up of a social barrier. It is not necessary to deny the historical significance and value of this verdict and this barrier in order to establish the fact that, in this way, the road to any further development is blocked in principle.

"A good end cannot be reached by bad means." This thought has always functioned as a roadblock in the way of new developments and new problems. This explains why there are so many problems at the moment more or less closely connected with the

[20] *Summa Theologiae* I–II, 18, 8; see *Tijdschrift voor Theologie* 4 (1964), p. 168.

field of medicine. And it is not only because the individual person is so concerned with these matters, but also because modern science has demonstrated its power of development most strongly in the field of medicine. One has only to think of such things as the rhythm method, especially in the years immediately following the discovery by Knaus and Ogino in 1929 and 1930, or transplantation, painless delivery, plastic surgery, the use of analgesics, birth regulation, euthanasia, brain surgery, reanimation, psychopharmacology, experimentation with living semen, artificial insemination, and so on.

On grounds of fundamental ethical considerations, and also in view of the recent past in the field of practical moral theology, it appears to me that we are neither exaggerating nor being unjust when we state that every approach to a moral problem in such terms as "bad means–good end" is suspect, to say the least. There can scarcely be doubt that attempts to solve problems, and especially new problems, will continue in this way. We repeat that, in a certain sense, this is a legitimate social procedure. But it is important to see its limitations, even if only to put the community on guard against a dictatorship of conformism. On the other hand, it is also true that a dictatorship of ethical relativism, which does not see or acknowledge the real significance of "social language," would make all normal community impossible.

The tradition of moral theology states that good and evil are specifically different. It states, in other words, that the human significance and content of action is always either communication or the frustration and destruction of communication. All concrete human action comes under one or the other of these two categories. Therefore, it is not the material content of the act which interests ethics in the first place. Ethics is concerned primarily with the deepest human significance of *all* action. It is concerned with community, with communication, or with their opposites.

In a Christian or theological ethic, this human evaluation takes on a much deeper significance. When ethics establishes that human action must be either good or bad, the theological ethic is reminded of the statement in scripture: "He who is not with me is against me, and he who does not gather with me scatters" (Lk. 11, 23).

There is no alternative. One is either with one's fellowman or against him—not according to one's intentions and aims, but in actual concrete fact. And if the outcome of the Judgment depends upon this (and, according to its description in Matthew 25, it does so depend) it can only be because the Judgment once again, and this time definitively, reveals the ultimate significance of the world and of human society. It reveals him who is both present and concealed in our world, who is both vulnerable and inviolable in it.

A theological ethic is not concerned exclusively with the question of good and evil. Part of its task is also to consider the subjects of merit and sin (*meritum* and *peccatum*). There is nothing to prevent our ascribing, in a certain sense, a supernatural significance and a meritorious character to human action. We are equally bound to acknowledge that we have a perfect right to speak about the sins which man commits and the forgiveness of sins. But it is obvious that a theological ethic must pose a question about the position of merit and of sin in the total reality of human action. It it sufficiently well known that the traditional moral theology regards fasting and abstinence (abstinence preferably to be understood in more than one sense) as meritorious, and unchasteness, on the other hand, as sin—to mention only one example.

Now, if along with this we remember what we have already established with regard to good and evil in general, it becomes immediately obvious that in the matter of merit and sin we have to deal with exactly those same phenomena observed in the human ethos. The ecclesial community manifests exactly the same traits and characteristics as any other community. Merit, sin, and indifferent act all serve in this community, too, as anterior qualifications of particular external acts. Once more we notice that there sometimes appears to be, as it were, a magical connection between the qualifications "merit" and "sin" and these external acts, just as this can be the case with "good" and "evil."

In other words, the generally accepted manner of speaking about merit and sin is characteristically pragmatic and directive, as is normal in every community, and up to a certain point it is quite justified. One may add that this once again emphasizes the pastoral task of the Church, for the Church is no less concerned for the salvation of the faithful than the civil community is for the welfare of its citizens.

There is an obvious parallel to be drawn. Both Church and civil community are controlled by the same sociological laws. Nevertheless, a theological reflection remains just as necessary, and, in a certain sense, just as necessarily must be contemporaneous as an ethic in regard to the ordinary "language" of social intercourse. A theological ethic must inquire into the sense and content of merit and sin—on the basis of the fundamental principles of anthropology, ethics, and theology.

In the light of all the above, we must come to the conclusion that good and merit coincide and that evil and sin coincide, without their being formally identical, however, and without their expressing precisely the same aspect. The People of God, or divine community, cannot come into being without human community, and human community, in fact, includes the divine community, whether or not it is actually recognized, although, of course, in the latter case it has no reality for man. This does not mean that the human community is the same as the divine community, because God is not totally defined by what is human, but he is present in man and manifests himself in man and in what is human. *Mutatis mutandis,* the same holds true of evil and sin. Sin exists only where evil exists, and all evil is, in fact, at the same time sin.

In other words, within the sphere of human action we are always and everywhere confronted with the God who chose to show himself and to be present among us in our own autonomous world. For this reason, a reflection upon human reality and its autonomy is a condition that must be fulfilled before we can understand the salvation proclaimed to us, which is the God who became man.

This is a question of more than mere understanding. Even though the significance of theology within the Church may well be relative, and, above all, even though each concrete action is infinitely more important than all theory and theorizing, the fact remains that reflection can sometimes help to unmask prejudices and mistaken views, thus freeing men from burdens they cannot carry and should not be asked to carry.

The "social language" is a necessary thing and a great benefit, but when it stifles and oppresses, the time has come for ethics, and particularly theological ethics, to fulfill its necessary task. By means of its fundamental reflections, ethics must make clear, and persist in making clear, that community and communication are the whole purpose of the social language, and that upon community and communication not only good and evil depend, but also salvation and damnation.

IV

Passion

In the previous three sections we have spoken of human action with reference to freedom, to intersubjectivity, and to good and evil. All that we have said describes hardly more than the skeleton of human action. It is an outline which is correct in principle, and it helps toward a thorough understanding of the main elements and the essential structures and significance of human action, but it remains an abstract outline and an elementary structure. The everyday reality is less transparent and seldom so clear. The dividing line between good and evil can be drawn very much less sharply in everyday reality than any theoretical analysis would tend to make us think. In daily life, good and evil appear rather to be confused and interwoven. Communication is achieved, but not without disruptive moments. There is no symphony in which we do not hear the strangest counterpoints for which we would seek in vain in the score.

Communication is sometimes gruff, sometimes unexpectedly smooth. Evening is often the best time for it; attempts are particularly unfortunate in the early morning. Moreover, there are innumerable varieties of physical condition that influence communication. Weariness, sleep, sickness, indisposition, food and drink, alcohol, tranquilizers, stimulants, vitamins, smog, a good night's rest—these are only a few of the numberless factors that, through the corporeity of man, impress their stamp on human exchange.

The idea that human emotions, dispositions, and disturbances (in the Latin tradition, *perturbationes, affectiones, affectus, passiones*) are all an attack on the human ideal of imperturbability of which the Stoics dreamed has no longer any serious following today, although it is a good thing occasionally to emphasize this fact. Nor are there any longer any serious defenders of the platonic form of dualism, but we are by no means far removed from this dualism, if only because of the way in which body and soul were spoken of in the recent past, and the way in which they are often spoken of even now.

Now, in this case, too, the theory must not be condemned over-severely. Even though the theories may seem to make no provision within their framework for the factor of passion, or corporeity, they actually do so provide. It is typical of practical morals, for instance, that corporeity is not considered worth a chapter to itself; it is thoroughly accounted for, however, in the social language and in the actual descriptions which we find in the manuals.

However, where the ethical tradition has reached a level deeper than that of the social language, we find that the factor of passion, or corporeity, was often investigated explicitly and sometimes at great length.

With this investigation we are now particularly concerned. In view of our previous discussions on corporeity and intersubjectivity, it is not necessary to argue all over again that corporeity exists entirely in function of intersubjectivity. It would also be superfluous to draw up a kind of general statement about the fundamental goodness of passion and corporeity, for it is only the actual concrete act of a man that can be qualified as good or evil. Even if an untimely pessimism has not blackened passion *a priori* —whether or not under the influence of a particular view of original sin—it must be remembered that rash optimism is also out of place.

It is very much to the point to remark explicitly, and in advance of any further discussion, that in this question we are entering a field which modern psychology has made particularly its own. This is a specialized field, and on entering it we need to define our competence as closely as possible. We do not plan to provide any overall view of the various opinions and methods one finds among psychologists. Nor do we plan to provide a summary of the most important conclusions of modern psychology. All we must do is indicate a number of things that are relevant to our own field, because they help us to understand the concrete shape in which freedom, intersubjectivity, and good and evil actually present themselves.

A number of physical postulates must be fulfilled if the corporeity of man is to be intersubjective and even communicative or successfully intersubjective. The classical ethical tradition has gathered a large number of observations on this point. If we inquire, for instance, into the sources of Thomas' treatise *"De passionibus animae,"* we encounter the following names: Aristotle, Augustine, Pseudo-Denis, John Damascene, Boetius, Nemesius,

Avicenna, Homer, Aristophanes, Cicero, Sallust, Vergil, Vegetius. Scripture, of course, comes in here. Nor should we omit the names of Seneca and Albert the Great (actually, Albert the German).

In this section we shall follow Thomas' treatise, partly because I am not competent in the specialized field of modern psychology, but chiefly because of the limited aim we have before us in this context and in this discussion, and because Thomas' treatise, to a certain extent, gives an inventory and summary of the entire tradition.

In this treatise, *"amor"* (love) is defined as a *"connaturalitas."* It is a connaturality, an adaptation (*"coaptatio"*), a fitting in with or accommodation, a self-accommodation to, a *"complacentia"* or a delight in something (someone). An "immutation" or even a corporeal "transmutation" has taken place: there is a material change which, in general, is more easily noticed in fact than described in detail. This change presupposes a "proportion" and a "similitude" between the persons involved: a certain equality and a certain preordination toward each other. And love itself leads to "union," unification, presence, unanimity, "mutual inhesion," reciprocal attachment, and even to "ecstasy" or, in a way, to stepping out of oneself. Following texts chiefly from the Song of Solomon, love is also described as "melting" as opposed to "freezing," as thawing and becoming tender so that the heart is open to the other.

Man is more than corporeity. Love is more than a corporeal, physical thing. But, in a certain sense, this "more" is nothing other than the physical reality. It is not something added over and above the physical reality. There is certainly not some sort of thing that exists apart from the physical reality. While it can happen that the physical reality does not grow into love and that it remains blocked, as it were, it is quite impossible for any love and friendship to exist which is not at the same time a physical reality in one way or another, more or less perfect, more or less integrated.

The tendency here is to move too quickly into the next chapter. We are still concerned simply with the corporeity of the concrete act; we are still concerned with a phenomenon for which conjugal love serves as *the* model almost in the very nature of things, and not only in the Song of Solomon, but it is a phenomenon which is, in fact, very much wider than conjugal love and which, in relation to intersubjectivity, can also be very negative.

"Fervor," a glowing warmth, is another factor that belongs to the sphere of this love. We can observe it most strikingly in children when they are enjoying the total surprise of a Christmas or birthday present or some other wholly unexpected happy event. When people are involved in an accident or a disaster, or when travelers see the last train pulling out and know that they will have an all-night wait for the next train, a comradeship begins to grow, a feeling of sharing the same fate that has all the marks of a rapid "thaw."

Love implies a "consonance," a kind of resonance. The opposite is the case with *odium* (hate) or aversion. There is a "dissonance," a lack of harmony, a difference in wave length, possibly even a repugnance stimulated by things that are described as being repugnant.

The question about the human significance of the physiological factors involved in aversion becomes much more pressing than in connection with feelings of well-being or pleasure. Although sympathy and harmony can be accepted calmly and without question, such is clearly not the case with aversion and repugnance. Thus, there is a sense in which these latter reactions reveal the subhuman character of all physical reactions and emotions. This is not to say or even suggest that these reactions are bad or indecent. We are simply saying that in *all* these feelings, without exception, corporeity is presenting itself as a task to be fulfilled in intersubjectivity and communication. Mindlessly to follow every pleasurable feeling can be just as wrong as simply to allow oneself to be overcome by aversion and repugnance.

These feelings, however, do occur, and they play their part in the intersubjective relations of man to man. This is the relevant point. It is further relevant that this whole complex of physical factors differs from one man to another, and that no two are exactly alike. Thus, corporeity is indeed a commission to be fulfilled in intersubjectivity; at the same time, however, strict uniformity and perfect synchronization of emotions and emotional resistance are impossible and inhuman because men differ in their corporeity. Thus again, the fact that corporeity is a commission given to each man to be fulfilled in intersubjectivity implies in no way that this commission is the same for every man or that its content will be the same for every man. On the contrary, each man discovers an entirely individual task in his own corporeity and in the tone of his own personal emotions; quite simply, there is no precedent, however canonized or holy, in which he can read off the details of his own task. In other people he can find a stimulus and, in a more general sense, an example to imitate. If he seeks more than this from an example, he is either overstraining himself or he is content to remain below the measure of his own capabilities. Of capabilities, and thus also of incapabilities, there is a great deal more to be said in this context, although in the nature of the case we shall have to confine ourselves to generalities.

Delectatio (pleasure, delight, satisfaction, joy) is connected with one or another form of activity and, in a certain sense, is the result of it. One can take pleasure in work itself—on condition that it does not overtax one's powers, in which case work becomes burdensome, displeases, and results in a greater need for rest and relaxation in order to overcome the *tristitia* (depression, despondency, exhaustion) brought on by this too prolonged or intense work.

The very nature of *delectatio* requires that it be taken in proper doses. This increases its stimulating effect; simple prolongation reduces it. Thus, just as with activity, a certain variation can produce joy and peace and satisfaction. "Variety is the spice of life."

There are, let us note, very many causes of and inducements to joy: hope and remembrance, the goodness of others, doing good to others, and so on. Young people and melancholics have the greatest need of pleasant stimuli, as the traditional wisdom of experience indicates. Young people need the stimuli because of the changes that take place in them as they grow; those who suffer from melancholy need them because of the despondency, the *tristitia* which they must try to overcome. Tradition has even brought out the very close connection between *delectatio* and *dilatatio*—*dilatatio* meaning the broadening, widening, clearing up, or airing that accompanies a pleasant stimulus.

One may ask whether there is not some sort of hedonistic philosophy behind all this. The answer is very matter-of-fact and decisive: *"Nullus* [*potest*] *vivere sine aliqua sensibili et corporali delectatione"*[21]—it is simply not possible to live without some pleasant experiences and some satisfaction of bodily desires. All this means simply that a reasonable and grateful acceptance of all

[21] *Summa Theologiae* I–II, 34, 1.

that corporeity implies is one of the essential requirements of intersubjectivity.

We see the same thing when we turn our attention to *dolor* or *tristitia*—suffering, pain, anguish, dislike, sadness, despondency, depression, and all related feelings and experiences. These tend to concentrate all attention on the self and to impede all vitality, and the more intense they are, the more this is so. They weigh on the mind and can lead even to extreme melancholy and mania. In spite of this, however, *tristitia*—difficulty, laboriousness—in moderate and reasonable doses can stimulate activity and intensify concentration.

Among the best remedies for feelings of displeasure are any kind of pleasurable feelings. In addition, one of the best ways to end dejection is to allow the tears to run their course or, in some other way, to give air to one's sadness. This not only reduces the inward tension ("Sorrow shared is sorrow halved"), but in the given situation it can itself be a pleasant sensation.

In a perhaps unexpected context, Augustine makes some highly realistic observations. Recalling in the *Confessions* the death of his mother Monica, he is clearly able to recall, many years after the event, not only that he was bitterly saddened and deeply struck by her death, but also that, after the company had gathered and prayed for his mother, and after long discussions with his friends in Cassiciacum, he decided to take a bath—this, after all, had a reputation among the Greeks for banishing depression. Although the bath did not have the desired result, his despondency was greatly reduced after he had a long sleep.[22]

Another factor that can play an important role in human action is that of *spes* and *desperatio*: hope, expectancy, optimism, and despair or discouragement. Hope finds its basis in whatever gives a man power to achieve what he intends to achieve. Hope can be founded on money, physical strength, experience, and so on, but it can also be founded on the illusion of these, while previous experience can deprive a man of all hope. This is the reason why old people often have very few expectations. We must also note, however, that optimism is often the result of stupidity or lack of experience.

"In iuvenibus et ebriosis abundat spes": optimism, the saying

[22] *Confessiones* IX, 12, 32; see St. Thomas, *Summa Theol.* I–II, 38, 5.

goes, is typical of the young and the tipsy. Of course, one does not want simply to equate the two groups, but there are points of comparison to be noted. Young people have hardly any past as yet; practically everything lies in the future for them. There is something inherent in their vitality and desire for adventure that others find supplied by the effects of alcohol. Young people have known very few disappointments, have experienced very few difficulties, and are inclined to think everything is possible. There are similar attitudes to be noticed in some people "under the influence"; they do not recognize dangers and are unaware of their own limitations.

For the rest, optimism is a weighty factor in every one of man's undertakings. It increases the intensity of his work: first by the conviction that he will be able to do what he sets out to do, then by the satisfaction that expectancy lends to the work itself. *Securitas* (self-sufficiency and self-assuredness) can do serious damage to optimism because it leads a man to ignore or to underestimate the difficulties before him.

Timor comes in many kinds: fear, anxiety, horror, timidity, shame, bashfulness. While death is what men must fear most, the remarkable fact is that men hardly ever fear death so long as it still seems far away, even when it presents itself as inevitable. The wisdom of experience seems to teach that we have a genuine terror of death only when death is very close with some possibility of escaping it.

The more suddenly and unexpectedly something happens, the greater the anxiety it occasions. There are several reasons for this. Unable to review the situation quite as quickly as it has arisen—at first sight it appears to be worse than it actually is—a person is unprepared for it. Experience, therefore, diminishes anxiety. Connected with this is the fact that quick-tempered people are feared less than the sly and the crafty.

Evil, however, is feared in the measure in which it threatens to endure. Accordingly, something is particularly feared as long as there is no known remedy for it—certain diseases, for example—for then there is no end in sight.

Particularly interesting in this context are the conspicuous physical manifestations of fear that tradition has observed: *contractio* (cramping together, shriveling, shrinking). A possible consequence of this (here, too, tradition is highly realistic) is that matter is ejected which in normal situations would leave the body in a more considerate manner. A racing heart, shudders, and shivering are further symptoms; still others are a trembling lower lip, a shaky jaw, trembling hands (which so easily drop things and knock things over), and knocking knees. Anxiety to this degree makes it impossible to carry on normally with one's work, but it is quite certain that just a little bit of anxiety can be a very good stimulant.

Audacia (audacity, excess of bravado) is, in a certain sense, the opposite of anxiety. The audacious man is aggressive in the face of evil or danger. At first sight this may appear to be an advantage, but it is actually a weakness. For when aggression has burned itself out, it is very likely that all strength will have been drained. If a man sets about things with greater care and consideration, he may not start off with such a spurt, but he will be able to bring all the more power to his perseverance because he took a better view of the whole task before starting.

Ira (anger, rage, temper, frenzy) is the last in a series of *passiones* from which we have been making a more or less arbitrary selection. Rage is said to have no opposite, whereas other passions are very often referred to in connection with opposites: love and hate, hope and despair, timidity and bravado, and so on. In a certain sense, rage can be described as the explosion of the conflict between hope, on the one side, and disappointment, on the other. Rage implies a certain deliberation and a measuring of the damage done against the revenge to be taken. For rage rises from some form of disadvantage or damage done to oneself or to some person with whom one stands in very close relation. To despise or disparage another is one of the ways of damaging a person, and it finds many different expressions, such as silence, disregard and forgetfulness—*"Oblivio parvipensionis est evidens signum"*[23] ("Forgetfulness is a clear sign of disparagement"). As Seneca and others have shown, this is something that was learned a very long time ago.

Since the chief cause of rage is disparagement, rage can spring from various sources. It can come from *excellentia* that is belittled. It can also come from the consciousness of a lack or a shortcoming, because people who are physically or morally deficient, or who have clearly fallen short in one way or another, are very touchy. On the other hand, those who have a genuine *excellentia* in an

[23] *Summa Theologiae* I–II, 47, 2 ad 3.

official, personal, scientific, or professional capacity are very much less touchy, because disparagement can do their real ability no harm, although it remains true that the more unworthy the manner of the disparagement, the more anger it is likely to incur.

An actual outburst of rage or an act of revenge brings a certain satisfaction, and this satisfaction lessens and dissipates the *tristitia,* the discontent that had helped to raise the temperature to the boiling point in the first place. Even the mere thought of revenge can have this calming influence to some extent.

The physical phenomena of rage or anger that have been observed in tradition are also of interest here. There is not only mention of a *"fervor sanguinis et spirituum circa cor"*—a heated concentration of blood and bodily humors in the region of the heart, according to the terms of archaic anatomy—but also descriptions such as we find in Gregory's *Moralia*: the body shudders, one trips over one's own tongue, the face becomes a fiery red, the eyes bulge, and so on.[24] One often says more in anger than it is good to say. But rage can also make one silent, either for fear of saying too much or simply because speech becomes quite impossible.

The descriptions we have given above are, in the main, quite simply a selection taken from the relevant passages in Thomas, which, to a great extent, are a summary of much older analyses and observations. Admittedly, this description could arbitrarily have been replaced by another, taken, for instance, from modern psychology and sociology. The main advantage of Thomas' treatise in this case was that in a favorable moment it saved us the labor of gathering new data.

Nevertheless, the essential object is probably achieved in this way, too. We needed to observe and reflect on man in his physical manifestations in order to complement the picture of man provided by the foregoing analysis of human action as (1) free, (2) intersubjective, and (3) good or evil. Freedom, intersubjectivity, communication, and the disruption of communication do not play out their roles apart from corporeity or above it, but entirely within it and on its foundation.

It would be quite out of place to begin now to speak prematurely of good and evil, and all the more so to speak of good and evil *a priori*. The entirely individual characteristics of cor-

[24] *Moralia* V, 45, 30; St. Thomas, *Summa Theologiae* I–II, 48, 2.

poreity have been incontrovertibly established by research into the biochemical structures of heredity. This corporeity, individual to each man, is the simple datum that must indeed become intersubjective, but which nevertheless in a certain sense remains itself and manifests a number of individual or general affinities, tendencies, repressions, resistances, and so on. These, in turn, may promote communication, or make it difficult, or even break it down.

To draw from all this the conclusion that freedom, intersubjectivity, and good or evil are, in fact, more vital, more impassioned, more emotional, warmer but also cooler, less sharply defined and less pronounced and, in a certain sense, less rosy than any provisional sketch would make us believe is but one thing, even if at the moment it is the most important.

We can move to another matter, and theoretical reflection will not be any the worse off for a consideration of it, and it is this. In human communication man must take into account many aspects, and there are still many aspects which escape him; and in taking into account and not taking into account, there are limits, so that when all is said and done, we can but wonder at the amazing fact of human understanding and human communication.

3. The Origins of Human Action

This third chapter is the last chapter in our treatment of the fundamental Christian ethic. It consists of two parts: the first considers man as the origin of his own action; the second considers God as the origin of human action.

The first part has three sections: (1) on *habitus* (character or basic attitude in general); (2) on *virtus* (authentic humanity); (3) on *vitium* (deficient humanity).

I

MAN AS THE ORIGIN OF HIS OWN ACTION

A. "HABITUS": CHARACTER, OR BASIC ATTITUDE IN GENERAL

Habitus plays a more important role in the great ethical and theological tradition than in the field of practical morals. It is clear from what we have already seen that this is natural and to be expected. Still, we should turn our attention, at least briefly, to *habitus* in order to see what aspect of the reality of man and of human action the tradition had in mind here.

The importance of a short reflection on habit will become still clearer if we understand what it was that led the theologians of the past to occupy themselves with *habitus*. Thus we come once more to the 12th and 13th centuries, whence our present-day concepts and formulations in theology still derive to a large extent.

The notion of the infusion of grace by baptism found a complement in the 12th century in the formulation: the infusion of the virtues. Studies, particularly on the baptism of young children, together with certain anthropological questions of a critical nature about the content of the concept of virtue (*virtus*), led the theologians of the 12th and 13th centuries to give attention to the problem of habit and of virtue in order to see how certain theological affirmations could be reconciled with relevant anthropological evidence. Not everyone could readily digest the notion that virtues were infused into little children, for habits and virtues in other contexts were certainly not defined as the kind of thing one would be inclined to look for in newborn babies.

With these remarks, we may leave the genesis of the problem. But it must be remembered that the concept of "infused virtue" contains a theological affirmation which still belongs to the *depositum fidei,* and rightly so, as we shall see. Thus, on the one hand it is meaningful to ask ourselves what our predecessors meant by these terms; on the other hand, it is surely realistic to take up a question which we ourselves may not be inclined to ask, but which others may ask at any moment, and in fact do ask, even if only with reference to some theological handbook which uses these particular terms, drawing on the heritage of the theology of the Middle Ages (and partly even on Augustine and other Fathers of the Church).

When we try to understand the reality which tradition was referring to when it spoke of habit, a number of words present themselves to an English-speaking person: character, attitude, basic attitude, mentality, and the like. Habit signifies that which is the source and origin of particular acts, and which, at the same time, is the result of concrete action. Habit also signifies an abstract reality in this sense, that it says nothing yet about good or bad, favorable or unfavorable. The same holds true of the English words mentioned: attitude, character, mentality—these still leave room for further specification in a favorable or unfavorable sense.

In the concept of habit, as in that of character and the others, we are once more faced with the human reality as entirely corporeal and entirely intersubjective.

The concept of habit includes the concept of a corporeity entirely in function of intersubjectivity. It includes the notion that this corporeity is directed in a definite way and even integrated to some extent, although the latter, in particular, is true only when applied to good habits, good character, and so on.

In other words, habit refers to "passion" when passion is no longer arbitrary and autonomous, but is, to a greater or lesser degree, an integrated part of the whole which we call a human person. Passion becomes habit—that is, it becomes freedom.

Habit, character, mentality, and the like thus express the fact that passion never occurs in a chemically pure state, but always, in one way or another and to a greater or lesser degree, in the form of habit, or character, or person. In the human sphere of the family, corporeity becomes intersubjectivity and communication, even if very gradually. If it did not sound so analytical and almost chemical, one would be able to say that "the baby's first smile" (Buytendijk) conclusively demonstrates the presence of habit. (Habit is a human reality; custom is a conditioning, a routine—useful and necessary, but something quite different from mentality, character, and so on.)

Without doubt, there is this difference between the Latin word "*habitus*" and the English word "character": *habitus* has a plural, whereas character, in this sense, does not. However, this seems to me to be a difference of a very secondary sort. In the last analysis, it is of no great importance whether the human reality should be indicated as a whole or according to its diverse facets.[25]

A question which is of importance, however, and of special importance for theology, refers to the manner in which Thomas explains the concept of "*habitus infusus*" (infused habit).[26] Habit, character, mentality, and so on are all brought into being gradually in one man's living in contact with other men. What are we to understand by the notion of habit infused into a man by God?

One possible interpretation mentioned by Thomas is that habit is constructed in a man by God "*absque ipsis causis secundis*"—without the intermediacy of any causes in the created order. But Thomas wishes to reject this interpretation. Before mentioning it, he gives another interpretation in which the created cause (*causa secunda*) is *not* excluded, although his choice of words here is so discreet that one might mistake the sense of this interpretation if one did not have the rejected one immediately following it.

Taking over a current form of theological argument, Thomas says that certain habits dispose a man toward a goal that exceeds his

[25] Cf. *Summa Theologiae* I–II, 54, 4: "*utrum unus habitus ex multis habitibus constituatur.*"

[26] Cf. *ibid.*, I–II, 51, 4.

nature. These habits must be in proportion to that goal and must themselves exceed man's nature. Therefore, these habits can only be infused by God.

As we shall see more clearly in the discussion that is to follow, Thomas does not find any contradiction between the notion of a habit brought into being by God and a habit that develops in the natural, human manner. In other words, just as the human situation remains in a certain sense unchangedly human but nevertheless must be called "beatitude," and just as the human act remains, in a certain sense, unchangedly human but must be called "merit" or "sin," so, too, habit remains in a certain sense unchangedly human but must nevertheless be called "infused." The reason for all this is that God himself is present in human form in man's own world, and God thus turns all that is human into a gift that is divine without its ceasing to be human.

In other words, infused habit does not signify a kind of reality discrete from and alongside of habits that are specifically human. The term "infused habit" expresses and reveals the character of specifically human habit as divine gift. This human habit is divine gift, but not in spite of its human character and human genesis; it is divine gift precisely because it is human and in that it is human, and not without the intermediacy of human causes.

Finally, and in general, we shall see that grace, too, is not a discrete reality alongside the human, but that it is entire man as divine gift, the "new creature."

B. "VIRTUS": AUTHENTIC HUMANITY

Habit is passion or corporeity that has become freedom, intersubjectively, and person. However, an actual man is not neutral in the way that habit is neutral. Each man in his actual mentality or intersubjectivity is either in accord with his fellowmen or in conflict with them. A man is either good or bad. In him, corporeity has become either virtue, which is authentic humanity, or vice, which is the lack of it.

This is the picture which an outline scheme provides. But here again, and even more so than in connection with particular acts, we must immediately notice that the reality is a great deal more complicated than the simple black and white of the scheme. No one divides the world into good men and bad. And if one should make such a division, it would never apply to people he actually knows. It is, therefore, all the more remarkable that we do allow ourselves to be swayed by such essentially untrue black-and-white pictures in a political or religious context. People accept such terms as "the Protestants," "the Russians," "the Chinese," "the Negroes," "the Germans," "the Jews." However interesting this may be, this is neither the time nor the place to ask critical questions about these incontrovertible facts and about the apparently universal tendency to extrapolate, project, and draw up neat schemes in black and white. The point with which we are here concerned is the spontaneous recognition and consideration of the fact that one cannot schematize and divide men into good and bad.

Nevertheless, the schematic division "virtue and vice" does occur and have a meaning. Its value is not based on any untrue picture of the reality in black and white, but rather upon an analysis of the tendencies in the reality. The destiny of a man's corporeity is communication, community, and the sharing of common humanity. These are not ephemeral and incidental manifestations of contact that is successful by happenstance. They are the fulfillment of a corporeity directed toward communication and which, in a certain sense, has become communication and solidarity. It is this fulfilled corporeity that we term "virtue," "authentic humanity," and the like; the absence of it we qualify as "vice," "lack of humanity," and so on.

The purpose of the division, then, is not in any sense to be able to point out how '"virtue" and "vice" are found in reality. On the contrary, the reflections from which this division emerges recognize clearly and concentrate upon the fundamental possibility of communication and humanity shared in common. This possibility is seen within the endless variety of men who are good and not so good, balanced and not balanced, pleasant and not so pleasant, good-willed and less so, men who, in fact, have achieved communication in part, have failed in achieving it in part, and still fall short of it. All this implies an entirely different vision and interpretation of the reality of man. There is no place in it for the principle formulated by Hobbes: *"Homo homini lupus"* (Man stalks his neighbor like a wolf). It does not agree at all with the view which declares hate, war, and conflict to be principles of human society.

The distinction between vice and virtue is not a criterion for the division of men; rather, it is a principle of interpretation which helps toward an understanding of man. It is an interpretation of a reality that one can describe as a conglomeration of "virtue" and "vice," of achieved "virtue" and "virtuous" design and opportunity that has been frustrated or has deviated.

Authentic humanity is achieved only gradually, and if it diminishes and declines, this, too, happens only gradually. There are deviations that are temporary and that pass, and there are deviations that are more fundamental. But not all the "vice" or lack of genuine humanity in the world can take away the fact that genuine humanity remains possible for every man and is the task of every man. It is a task that every man is able to fulfill in his intersubjectivity, and, therefore, partly in dependence on his fellowmen. This is more or less what the traditional term "virtue" intends and what is signified by the distinction between "virtue" and "vice." The only permanent source of communication and community is corporeity that has become authentic humanity.

Three points in the traditional treatment of the question call for special attention: (1) the distinction between the divine and the cardinal (or moral) virtues; (2) the concept of *"virtus infusa,"* or infused virtue; (3) the gifts of the Holy Spirit.

1. *Divine and Cardinal Virtues*

The distinction between the *divine* and *cardinal* virtues is the subject of a chapter in Part III of our study. However, a brief explanation is needed here.

Throughout the tradition, the trio of divine, or theological, or theologal virtues—faith, hope, and love—has remained constant. It derives quite simply and directly from St. Paul, particularly in 1 Thessalonians 1, 3 and in the much more familiar 1 Corinthians 13, 13. The four virtues called "cardinal"—thoughtfulness (prudence), fair dealing (justice), courage (fortitude), and moderation (temperance)—are also called the "moral virtues." They are found already in Plato and remain constant as a foursome throughout the entire classical and Christian ethical traditions. The most relevant fact is not that three virtues are distinguished on the one hand and four on the other; to a certain extent this is quite incidental. The essentially relevant fact is that distinction is made between the virtues called "divine" on the one hand and those called "cardinal" or "moral" on the other. In a particular way this distinction expresses the very core and essence of the Christian message. It is precisely in this respect that Augustine, Thomas, and many others are in fundamental agreement. Already in Augustine's day these distinctions were traditionally used to bring out and stress once again that human community and divine community, common humanity and the relationship to God, are materially identical but formally distinct. We have already seen this in the chapter on Christ, in the chapter on the meaning of human life, and in the third section of the chapter that followed, on good and evil in human action.

The same reality—that is, the same authentic humanity—is qualified, from one aspect and even primarily, as "divine virtue," while from another aspect it is qualified as "moral virtue." The reason for this is God's human presence in man and in man's world. From this it follows that authentic humanity, from one aspect and even primarily, is a *"consortium divinae naturae,"* a partaking in the divine nature,[27] an association not indirect but direct, with God who has become man. From another aspect, however, this authentic humanity is, purely and simply, humanity shared in common with one's fellowmen.

We found this same twofold reality in the meaning of human life which is, in fact, communion with God in the human community and by means of it. We found it also in human action, which is "merit" or "sin" inasmuch as it is good action or bad action with regard to the community and to one's fellowman. We find it here again on the level of "virtue." Authentic humanity is essentially and primarily a partaking in God's own life in and through our humanity shared in common with our fellowmen. Quite apart from any particular terms or schemes of division, this shows that the reality of the distinction between divine and cardinal virtues is an essential factor in a Christian ethic.

2. *Infused Virtues*

The concept of *"virtus infusa"* (infused virtue) is in a sense, already explained in our explanation of *"habitus infusus,"* or infused habit. What we say of habit must, in the very nature of the case, also be said of virtue, not only because "habit" is only an

[27] 2 Pet. 1, 4; cf. St. Thomas, *ibid.,* I–II, 62, 1.

abstract term for a reality that must be either "virtue" or "vice," but also because here, too, the same considerations are valid. The primary and essential reason for speaking of an "infused virtue" is the affirmation of the *causa prima,* or of God's good work in and through creaturely causes, and not in any sense a denial of the *causa secunda,* or the created and human reality.

The term ""infused virtue" is, therefore, not a denial of human reality and the human process of growth, but an unveiling of its character as gift and as divine.

How does this notion of "being infused" relate to the reality which, in a way, is a twofold reality, of divine and moral virtue?

The above consideration of the distinction between divine and moral virtues as two real aspects of one and the same human reality can help us to state this problem more clearly. The question to be asked in the first instance is not whether both divine and moral virtues are infused; rather, we must ask what is the meaning of infusion by God, as applied to the one human reality, qualified, in one aspect, as divine, and, in another, as moral virtue. Along with this, we remind ourselves that in a certain sense our investigation begins with and is concerned primarily with moral virtue and that the reason for speaking about the divine virtues lies in the revelation of the divine dimensions of this cardinal or moral virtue.

Of all meanings that the phrase "infused virtue" can have, the first is this. Moral virtue, or authentic humanity, is truly enough a human reality that comes into being in and by means of communication among men, but, at the same time, it must be and rightly is qualified as something infused in man by God, inasmuch as the entire human reality is brought into being and maintained by the creative power of God. In other words, because God is the creator, or the *causa prima,* and, in fact, the most fundamental cause of all that human causality is able to bring about, moral virtue, too, or authentic humanity, must be acknowledged to be a gift from God—that is, it, too, is "infused."[28] Therefore, when the theological tradition speaks of infused virtue, the first thing to be understood by this term is that all authentic humanity is actually a divine gift.

Consequently, the fact not only of God's creative activity in everything that is active, but also of his human presence in man

[28] Cf. *ibid.,* I–II, 63, 3, and 62, 1.

and in the world, means that human life is a constant and direct association with God himself. In other words, authentic humanity, or virtue, is called forth, not only by pure and simple humanity, but by God himself who is so truly manifest in one's neighbor, so that the tradition, together with scripture, can speak in such terms as *"frui Deo"* and *"videre Deum."* In its turn, and partly for this reason, authentic humanity itself is not pure and simple humanity, but is so really a manifestation of God that the expression "partaking in God's own nature" has always been applied to it.[29]

This unveiling of man and the human community in its aspect of mystery (and, therefore, the unveiling of God's presence) is expressed by the theological tradition in terms such as *"virtutes theologicae"* or "divine virtues." These terms, in fact, indicate the same reality as that which was first qualified as authentic humanity. However, in the precise aspect referred to by the term "theological virtue," there can be no question of being infused in one sense and of being acquired in another. Virtue, in this respect, is pure divine gift—that is, simply infused. For, precisely in that aspect of human reality by which it is God's own presence, it can be regarded only as gift and not as something that man acquires. Insofar as human reality is something that man achieves, it does not reach beyond the human level. The fact that it is, nevertheless, something that reaches infinitely beyond the human level can be attributed only to God's human presence in man and in the world.

Therefore, in the case of the divine virtues, "infusion" is not merely one of the aspects of the human reality as it is in the case of the moral virtues, because, from another aspect, these moral virtues are *acquisitae* (what that man himself acquires). Divine virtues express the pure gratuity of the divine gift, a gift which is not called for by any human need or by any human asking. Moral virtue must be called "infused" in one particular sense; theological virtue cannot be other than infused.

3. *The Gifts of the Holy Spirit*

Finally, we must say something about another traditional element that belongs in this context: *the gifts of the Holy Spirit.*

The Hebrew and Syriac texts of Isaiah 11 mention only six "spirits." Nevertheless, the entire patristic tradition, including the

[29] Cf. 2 Pet. 1, 4; also St. Thomas, *ibid.*, I–II, 62, 1.

Syriac, speaks of the seven gifts of the Spirit. The origin of this terminology lies not only in the Greek text of the Septuagint, but also in the number 7 as the scriptural (Semitic) symbol of perfection. The perfection of the gift of the Spirit is, therefore, the essential theme in the whole patristic tradition. Practically none of the Fathers apart from Ambrose took pains to attach separate meanings to the seven gifts of the Spirit. Scholastic and later theology very often did take such pains for various reasons. There was fascination in the possibility of relating the seven gifts to the three theological and four cardinal virtues, or to the seven sacraments, or to the seven capital sins, or the seven petitions of the Our Father, and so on. Nevertheless, like Thomas and like a fundamentally benign interpretation of the later tradition, we find here a resurgence of the central patristic theme: the fullness of the gift of the Spirit.

Does this bring a new complication into our "supernatural structure"? Not in the sense of this question. No new entities are added to virtue, or to authentic humanity, even in its character as mystery. The affirmation of the gifts of the Spirit is the affirmation of the activity of God's Spirit in and through authentic humanity, with particular reference to the divine community realized within authentic humanity. Thus, for example, when Thomas says that the seven gifts are dispositions rendering man sensitive to the workings of the Spirit, he does not envisage any new entities, but refers to authentic humanity itself as that which is produced by God himself and as that in which he works, for, indeed, he gives his Spirit without measure.[30]

In the first two sections of this chapter, our subject has been man as the origin of his own actions. In actual fact we have almost always been speaking about God as the origin of man's actions, compelled as we were by much of our data and many expressions taken from the theological tradition.

Naturally, there is no reason why we should not see that God is, in reality, at the source of all human activity and even at the source

[30] In connection with the number 7 as indicating perfection, it is interesting to note that in the debates which preceded the dogmatic definition by the Council of Trent of the seven sacraments no *argument* is brought forward other than the patristic reflection on the fullness of the gift of the Spirit (apart from some references to the Council of Florence and to the Scholastics).

of the human source of all human activity. At the same time, if we wish to conduct a human reflection and to justify and give an account of the conviction that is alive in us, we may not neglect or ignore the human reality. Corporeity and passion are intrinsically directed toward becoming genuine humanity and, therefore, toward becoming the source and origin of communication and community. This communication and community, thanks to God's human presence, becomes "beatitude," divine community, and Church. We can say all this the other way around. Through his manifestation in human form and through the working of his Spirit, God is so completely the origin of our activity that he makes us ourselves and our own humanity its source and its origin.

In agreement with the ethical tradition, therefore, the autonomy of man is once again recognized and the entire theological tradition is reaffirmed in the witness which it bears to the mystery, the mystery of Christ, the mystery of God's presence in humanity.

C. "VITIUM": DEFICIENT HUMANITY

Evil indicates the privation of good, or a deficiency in it. With regard to human action, evil indicates a lack of or a deficiency in communication with man and with God as well. In the same way, *vitium* (vice, human deficiency) indicates a lack of or a deficiency in *virtus* (virtue, authentic humanity). Human intersubjectivity leaves open no possibility of a third choice. A man is either directed toward others and toward community or he is not. Passion and corporeity do not exist in a state of neutrality for the simple reason that they do not exist in a state in which intersubjectivity would not apply. Passion and corporeity represent an inherent inclination and directedness toward others and toward community. The only alternative to actually being thus directed is deviation. In other words, the terms *"vitium,"* "vice," "human deficiency," and "weakness" all refer to what is still lacking to "virtue," or authentic humanity.

This deficiency of which we are speaking is qualitative rather than quantitative. It is not to any great extent the kind of deficiency we are thinking of when we say that someone has certain vices or shortcomings. These are, in fact, merely the symptoms of a weakness, a deficiency, a deviation all along the line. They are symptoms of a lack or incompleteness of direction toward others and toward community. It may well happen that we may notice this in the particular behavior of an individual in a concrete situation: for example, a lack of flexibility, a rushing after the illusions of one's own ideas, a lack of a sense of humor, and so on. We can observe these things because we often do meet people in peculiar situations and even more because the individuality of every man has its own characteristics and unusual traits resulting from physiology, education, status, occupation, and so on. In addition, the things that are noticeable in a man's character are also determined in part by the individuality of the observer. One will be more sensitive to honesty and dishonesty, another to flexibility and rigidity, a third to fairness in the exercise of authority, the misuse of power, and so on.

We do have a certain right, therefore, to speak about vices and deficiencies, but the fundamental meaning of this kind of talk does not have much to do with the determined and, in a certain sense, quantitative measure of this or that deficiency. Fundamentally, it refers rather to a qualitative deficiency in the state of being directed toward others and toward community, a qualitative deficiency which I can experience in a variety of particular ways in a variety of particular contexts. When I experience a vice or deficiency in another and speak out about it, I am always, at the same time, saying something about my own individuality as well, and not merely about the individuality of the other. On these grounds, therefore, the things that are actually said about vices and deficiencies can very readily be explained (of course, the same is true of the things said about virtues), and it is also easy to explain why different people can judge differently and point to different weaknesses.

One man will experience and formulate vice in one way, and another will do the same in another way. There are many expressions from many points of view, but a qualitative deficiency in the state of being directed to others and to community is the fundamental significance common to them all. Vice, lack of virtue, human deficiency, or weakness (and, incidentally, virtue as well) are all seen to be indications of a much more fundamental reality than appears at first sight. They refer not merely to a surface vice or deficiency, but, as it were, to a broader and more deeply seated shortcoming in directedness toward others and toward community.

In connection with this human reality, three theological questions are specially relevant. We shall deal with them at once.

1. *Sin*

It is obvious that the shortcoming just mentioned comes to light and finds expression in concrete action. Thus, a great deal more than the particular action alone is involved when we deal with bad action, and especially with bad action in the sense of "sin." Man reveals himself in his action. In deficient action, the man reveals himself as deficient.

Concrete acts are very important in mutual contacts between men. But the decisive importance belongs not so much to the acts as to the man himself as the source of that action. The fact that sin is spoken of as an act must not blind us to the deficiency in directedness toward God and toward fellowmen which, in a certain sense, is much more important than the particular acts in which this deficiency is brought clearly to light. In the particular act, a man is concretely confronted with what he really is, but this confrontation is a passing one. Evil and sin are not the kind of thing that passes, but are inherent in man, qualify him intrinsically, and first bring out the full significance of the name "sinner."

Nevertheless, one can no more sort men out into the sinners and the just than one can sort them out into the good and the bad. It is the sinner who is called the just man, and the just man who must be called a sinner. The French have a wonderful expression: *"les défauts de la qualité."* This is more than a diplomatic way of pointing out a person's faults. It is a very positive and deeply discerning manner of saying what most men, in fact, are, and this, too, in the light of the Gospel message.

Ecumenism is leading us to speak more often in terms of *"simul iustus et peccator"* ("man is, at the same time, both sinner and just"). In the sense of which we have been speaking, this is certainly a good thing, if (to express it in formal terms) the phrase "at the same time" refers to man and not to the *habitus*. For if the phrase is taken to refer to habits, we no longer have a paradox, but a real contradiction.

In fact, however, the adage *"simul iustus et peccator"* implies much more. On this point, however, we can be clear—clearer than the ecumenical dialogue sometimes appears to be—only after we have examined the next point closely.

2. *Original Sin*

In recent years the dogma of original sin has been the subject of many discussions and publications. There was every reason for this. From the camp of biblical exegesis came the announcement that the story of Adam and Eve "did not really happen." This destroyed the historical foundation of a sin that all men had inherited from Adam, and it was not long before questions began to be asked about original sin itself. No Adam, no Adam's sin, no original sin; it seemed fairly conclusive.

There was another reason for the recent interest in original sin, and it was an equally important one. The theory of evolution, with or without the implications and suggestions introduced into it by Teilhard de Chardin, meant the complete collapse of the bucolic image of creation and of the world that, up until that moment, had preserved undisturbed the *mise en scène* for a paradise of happiness and the sin of Adam. Should one still go searching among the ruins for original sin?

The dogma of original sin does belong to a very explicit tradition in the Church. Prescinding from the work of two lesser councils—the Council of Carthage in 418 and the Second Council of Orange in 529—the Council of Trent devoted to the subject a special session with "chapter" and "canons." Over against all this, we have the old story of the gardener who, because of a little theft, was dismissed and sent away with all his children by the lord of the manor. In our earliest catechism classes we all heard one version or another of the story, none of which appealed very much to our sense of fair play. At any rate the future for original sin did not seem very rosy.

The year 1950 came, and August, the month when everybody is on vacation, or ought to be. The bureaus of the Roman curia were closed; the Pope was at Castel Gandolfo with its restful views of the blue waters of Lake Albano; everyone who was able had fled from the suffocating heat of the Roman summer; all was profoundly at rest. But on Saturday, August 12, the Church was shocked into wakefulness by the most unexpected and violent storm in living memory. Pius XII had published his encyclical *Humani generis*.

From the Netherlands the storm seemed to be fairly distant. At that time we were not particularly progressive. But the storm struck France with devastating force. The chief victim was the *nouvelle théologie*. Ecumenical "irenism" and "dogmatic relativism" were condemned in the strongest terms, and very many dangerous tendencies were named. Now, seventeen years later, after the close of Vatican Council II, one feels inclined to say that *Humani generis* was one of the last violent and desperate sorties undertaken by the defenders of the Church of the medieval establishment, the "bulwark" against the builders of the new and open city of God on earth.

Nor were the defenders altogether wrong. *Humani generis* warned against too easy an agreement with the new theories about evolution, polygenesis, and the meaning of the first chapters of Genesis. This gave support to the dogma of original sin which was being threatened. It did not die quietly and without fame among the ruins of the ancient city. Theology was spurred on to renew the search for the meaning of the dogma of original sin, now that the foundations seemed to have been blown from under the traditional explanation.

From our earliest picture-book days, we have been familiar with the sad and exciting story of the Fall—Adam and Eve driven forever from paradise, with sanctifying grace, immortality, and all the other supernatural gifts taken away from them. Their lot determined the lot of all the children of men. By baptism, original sin is "forgiven."[31] This is why the newborn baby is taken to the church on the day it is born, or as soon thereafter as possible, and baptized with the water that was blessed on Holy Saturday and kept very carefully ever since. Mother sends a little demon to church, and a little angel comes back.

If there was any fear for the life of the child, not after but before the baptism we have just described, the child received something called "emergency baptism." If there were any serious difficulties even before its birth, one had recourse to something called "intra-uterine baptism," accomplished with a sort of syringe. A declaration from the Holy Office, dated August 21, 1901, in answer to a query from the Archbishop of Utrecht, permitted, in order to prevent infection, the addition of one part per thousand of chloride

[31] *Catechism* (den Bosch) (1948), q. 230.

of mercury to the water used in such cases. If a woman died while she was pregnant, a section had to be performed to enable the fetus to be baptized. Missionaries, equipped with this sort of moral (!) theory, went to China. Later, they were to write to Rome to say that the Chinese found such practices utterly repugnant. The Holy Office replied, on February 15, 1780, telling the missionaries that, in spite of the desirability of the practices, they were to refrain from any action in future cases. However, they were to try to bring the people to a better understanding and to help spread a knowledge of surgery. There are very many instances and decrees of a similar sort. They all help to make it incontestably clear that belief in original sin and in its forgiveness through baptism belongs to the tradition of the Church.

Names connected with the new investigations into original sin are chiefly those of Schoonenberg, Hulsbosch, and, more recently, Trooster and Haag. Their interpretation, in the main, is characterized by the phrase "situation of non-salvation." Schoonenberg was the first to use the term, and he supports it by reference to an expression in the works of Karl Rahner.

According to this interpretation, one can understand original sin as the salvationless condition in which a person comes into the world, because he comes into a world that is marked by the sins and shortcomings of men. These sins and shortcomings have been present from the first beginnings of the human race, no matter how imperfect and primitive the first men may have been. Therefore, nobody is free of original sin, and in this the interpretation is in agreement with the traditional teaching of the Church. Nevertheless, original sin in this interpretation is not merely something extrinsic, but rather something completely and profoundly personal, and this is in accord with man's essential intersubjectivity, in virtue of which a man is inwardly specified by the situation surrounding him. In this sense, original sin is incurred through birth, although not precisely through generation and conception.

This is only a very general description of the interpretation, but it will suffice here. Very likely, it does not do full justice to the authors concerned, particularly to Schoonenberg who has examined and defined the implications of his view very precisely in a number of publications. However, this is beside the present point. For we are not now involved in a dispute with the authors mentioned, but wish simply to give a short summary of a modern interpretation of original sin which is already enjoying fairly wide acceptance, more or less in the form given above, and often under the title of the "situation of non-salvation."

It seems to me that this interpretation seriously falls short as a representation of all that the Church's tradition understands by original sin. Let me add that I do not yet even want to pose the question of how Christ would be free from original sin or of how Mary could be conceived immaculate in a salvationless situation such as the one described, because these points involve a number of other problems quite different from our present one. Nevertheless, as far as I can see, these two questions alone would prove an insuperable obstacle to the theory of this interpretation. Vatican Council I very rightly stressed the principle of the *"nexus mysteriorum inter se et cum fine hominis ultimo"* ("the coherence of the mysteries of the faith among themselves and with the ultimate destiny of man). This new theory of original sin pays very little or no attention to the coherence of the mysteries of the faith.

A proper point of departure is the first thing we need if we want to try to understand the dogma of original sin in a way that will accord with the entire tradition of the Church. This is so, particularly since a too profane/historical interpretation of Genesis has fallen away. With regard to original sin, the starting point is of the same kind as with regard to sin. Sin is not simply a given reality which theology must try to describe and explain. The most elementary theological approach to sin establishes that the expression "sin" is an *interpretation* of the human reality in the light of the message of salvation.

In the same way, a theological statement about original sin cannot take original sin simply as a given fact and then proceed to attach some meaning to it, more or less under the guidance of what tradition has to say about it. On the contrary, the first thing that must be established is that the term "original sin" is an *interpretation* of the human reality in the light of the message of salvation, an interpretation of human reality that must be made because of the message of salvation. In other words, original sin is not some kind of discrete earthly reality which one must be able to observe and define or else it will have no meaning. Original sin is an essential part of the message of salvation itself.

Original sin is much more than the situation into which a man comes when he enters the world, a salvationless situation marked by the sins of men. In a certain sense, original sin is something entirely different from this situation. Original sin is the total human situation considered in itself as such, the pure and simple human situation, *in comparison with* the divine community and the situation of divine salvation toward which man is destined and called.

Original sin, therefore, is not something invented by Augustine, as the historian, J. Gross, thinks he can maintain in his study on original sin, a work which is as prejudiced as it is long. It is true that Augustine was the first to use the term *"peccatum originale"* and that he used it very frequently. It is also true that the Council of Carthage in 418, on the matter of original sin, was convened and drew up its formulas entirely under Augustine's inspiration. It is true that the inspiration for the Second Council of Orange in 529 was entirely that of the Council of Carthage and of Augustine. It is further true that the Council of Trent intended, in principle, to do no more with regard to original sin than simply to repeat and affirm the tradition. To a large extent, it did this quite literally, in terms borrowed from Carthage and Orange and, therefore, from Augustine. But in a certain sense, through the term *"peccatum originale,"* Augustine had done no more than formulate the very essence of the Christian proclamation, recognized and acknowledged by the whole of tradition in the numberless words and images of scripture.

Practically all of Augustine's many writings which deal wholly or in part with original sin (hereditary sin, primordial sin) are directed against Pelagius and Pelagianism, from the first friendly but clear letter that Pelagius received in the autumn of 410 (when he had just arrived in Africa after fleeing from a Rome threatened by barbarians) to the last great books against Julian of Eclanum and against the semi-Pelagian monks in the south of Gaul. It was there, only 100 years after the death of Augustine, that the Second Council of Orange, more or less definitively, put a stop to the tendencies to hollow out, caricature, and destroy the Christian message, tendencies against which Augustine had fought with unprecedented vigor. This, more than anything else, won for Augustine the title "Doctor Gratiae" in the Western Church.

What was it that so preoccupied Augustine? What made him so violent—at first sight, incredibly violent—in his polemic against Pelagius and his followers? What depends so much on the fate of the dogma of original sin?

Pelagius thought that baptism was not necessary for children or, in any case, that such baptism did not mean a forgiveness of sin, for children could not have sinned. Pelagius would have nothing to do with the notion either of original sin or of the sinfulness of man resulting from the primordial sin of Adam. Men sinned *as* Adam sinned, but they did not sin in some mysterious way *in* Adam, as Paul indicates or suggests in Romans 5, 12, a text very often quoted by Augustine. For Pelagius, the key concept of the controversy was sin, *"imitatione, non generatione"* ("everyone follows Adam's example in sinning, but no one inherits his sin"). The position of his opponent was precisely the contrary: *"generatione, non imitatione."* Pelagius believed in the goodness of man, a goodness which the creator himself had noticed, as the Genesis narrative and other texts tell us. On the other points, too, Pelagius gives the impression of being sincerely convinced that it is derogatory to the creative work of God to look upon man as fundamentally held captive and dominated by sin.

On the opposite side of the controversy, the central concept in Augustine's argument was not, in the first place, the necessity of baptism for children, nor was it an affirmation of the doctrine of original sin; his central concept was the fact that then Christ would have died to no purpose (Gal. 2, 21). Thomas, 850 years later, would recall precisely this same argument in his explanation of original sin.[32] Augustine's affirmation and preaching of original sin subserve his proclamation of redemption by Christ, and it is, so to speak, the opposite pole or the photographic negative of redemption by Christ. This alone makes it possible to understand what inspired Augustine both in his use of the term *"peccatum originale"* and in his emphatic affirmation of the reality of this sin.

If Pelagius denies original sin, then he denies grace, and he must deny the mystery of Christ and redemption by Christ. If Pelagius denies original sin, then he must deny the whole of scripture which sees in the human situation nothing else than blindness, deafness, lethargy, shadow, obscurity, dryness, dead bones, slavery, exile, and death, not as if every man were blind and deaf and so

[32] *Summa Theologiae* I–II, 81, 3: ". . . *alioquin non omnes indigerent redemptione quae est per Christum.*" For a more thorough investigation of the position of Augustine and the interpretation of Thomas, see the first part of my article, "Toward a Renewal of the Theology of Marriage," in *The Thomist* 30 (1966), pp. 307–342.

on, but because men are nothing other than blind and deaf and so on *in comparison with* the light, the sight, the sound, the abundance which no eye has seen, nor ear heard, nor has it entered into the heart of man to imagine, and, yet, which God has prepared for men. (For this last reason we may not say, therefore, that God has created man in sin.)

There were so many things that Pelagius, in his naive optimism, did not notice. When one spoke of original sin, of the power of sin, and of the slavery of sin, Pelagius did not see that this was not a matter of an imagined *moral* depravity in man, but that it all referred to being without salvation, to the slavery in Egypt and the captivity in Babylon, to being far from the promised land and the holy city of God. He did not see that it was all a way of speaking of the absence of the only One who would actually be able to give meaning and content to this sublunary existence. Nor did Pelagius see that it was all a description of a life that was really death as long as he had not changed it all to life. For man in his own autonomy, this death, this being without salvation, was the original sin of which Augustine and the whole of the Catholic tradition had been speaking.

From these thoughts we are at once carried forward into all the richness of the scriptural proclamation of God's redeeming presence in our world. Death, obscurity, exile, slavery, original sin—all these constitute the absence of God. Where God is, original sin is no longer.

The message and the sacramental proclamation of the message reveal this presence of God in man and in the world. This is why the tradition calls baptism an "*illuminatio*" (Gr., *phootismos*), a lighting up of the human darkness. For baptism, indeed, is certainly not a chemical cleaning process, but a "sacrament of faith," and, with regard to little children, it takes its power of significance in a certain sense from the faith of the ecclesial community. This is to say that the baptism of little children is primarily an act of the faith of the Church which acknowledges and exercises not only a human care, but, before and above all, a divine care for every child of man born into the darkness of man's world. Thus, from the very beginning, human goodness and care are established as sign and reality of God's own personal care for each individual man.

We believe in a baptism "for the remission of sins" for all men alike, including little children (as we profess in the Creed). Both Augustine and the Council of Carthage expressed emphatically that this faith in no way whatever could imply that little children had already committed any personal sin. When tradition, nevertheless, refers to the *"culpa,"* or guilt, of original sin, it is not in opposition to Augustine and Carthage; on the contrary, along with them and with the Council of Trent, tradition affirms that original sin touches every man and is personal to every man, even though it is not a personal act.

In tradition, original sin is the lack of "sanctifying grace" or of "original justice"—this last, with reference to the story of the Fall. This original sin belongs to man as such, to man as man. This is the sense of all the affirmations about descent from Adam and the transmission of original sin through human generation. This is the meaning of original sin despite a number of somewhat superficial complications which the later stages of tradition rather naively attribute to Augustine. This is the sense of all the statements about the children of Adam—which means "children of man"—in scripture. It is more than clear that this is the sense in which Thomas understands this sin of Adam. For example, in one and the same article and with a solemnity that is quite unusual for him, Thomas affirms that the Catholic faith teaches that Adam's sin is transmitted to all men; immediately after this, Thomas states that an inherited defect cannot be a sin; he then continues with a quotation, naming Aristotle as the author, to say that one cannot blame, but only sympathize with, a man born blind. (This is the briefest possible reference to John 9 and a summary of the whole chapter, especially if we remember Christ's own words: "It was not that this man sinned, or his parents, but that the works of God might be made manifest in him.") Thomas immediately follows this with a quotation from Porphyry to the effect that all men are, in a certain sense, one man, one Adam, because they all participate in the *"species"* (they are all included in the abstract idea) of man, although of course, Thomas himself is not an idealist in the tradition of Plato.[33]

The notions of heredity and the sin of Adam have nothing to do, therefore, with complicated theories about corporate personality, or about the juridical or physical inclusion of all men in the first

[33] *Ibid.*, I–II, 81, 1.

man Adam, or any similar theories which some maintain they have derived from Thomas' article of which we have just spoken. The salvationless state is the actual condition of man *as man,* as the "child of Adam." In this sense it is possible to speak of original sin as *hereditary,* simply because the children of Adam beget children of Adam, which is to say that man begets man, whereas God alone begets the children of God.

But does God's begetting, then, differ from and, as it were, compete with man's begetting? The answer, in a certain sense, is no. The children of man are really children of God, just as Adam, throughout the Old Testament and into the New Testament, is called "the son of God" (Lk. 3, 38). But just as we cannot say that authentic humanity is of itself divine (cf. our conclusion under "Infused Virtues")—in other words, just as we cannot say that the children of men as such are children of God—neither can we say that men as such bring forth children of God. This is reserved to God alone. How can it be otherwise? But God does make men do *more* than they themselves can imagine or comprehend. And here, once more, we must remember that that which was from the very beginning of the world must once again be made manifest and be seen—again and anew for each and every individual man. In our world, God remains absent and hidden as long as his presence is not proclaimed and made present to man in the proclamation of sacrament.

Scripture, as we know, has its own particularly clear and striking ways of expressing and proclaiming that the children of God are born of God. They are brought into the world by the virgin and by the unfruitful woman, begotten "not of blood nor of the will of the flesh nor of the will of man, but of God," as the prologue to St. John's Gospel (1, 13) puts it.

In the book of Genesis we are assured six times that Abraham's wife Sarah was barren,[34] and precisely because she was barren, Isaac, the son of God's promise, was born of her: "I will bless her, and she shall be a mother of nations" (Gen. 17, 16). Later we read: "The Lord visited Sarah as he had said, and the Lord did to Sarah as he had promised. And Sarah conceived, and bore Abraham a son. . . ." (Gen. 21, 1. 2). This is the first but by no means the last time that scripture speaks of God's visiting a woman.

[34] Gen. 11, 30; 15; 16; 17; 18; 21.

Isaac's wife Rebekah also was barren. "And Isaac prayed to the Lord for his wife, because she was barren; and the Lord granted his prayer, and Rebekah his wife conceived" (Gen. 25, 21).

Jacob, the son of Isaac and Rebekah, married Rachel, "but Rachel was barren" (Gen. 29, 31) until "God remembered Rachel, and God hearkened to her and opened her womb. She conceived and bore a son . . . and she called his name Joseph" (Gen. 30, 22–24).

Abraham, Isaac, and Jacob, the patriarchs of the nation, are not the fathers of the People of God. Moses would indict this nation: "You were unmindful of the Rock that begot you, and you forget the God who gave you birth" (Deut. 32, 18). God alone is the Father of his people and of the Son he begot, and the whole of the Old and New Testaments bears witness to this.[35]

The theme of virginity, of barrenness, and of birth from God in the Old and New Testaments is sufficiently well known, so that only a few select references will suffice. The barren woman is visited by God and rejoices (Is. 54, 1; Gal. 4, 27); Samson's mother is visited by the angel: "Behold, you are barren . . . but you shall conceive and bear a son" (Jg. 13, 3); the birth of Samuel (1 Sam. 1); Isaiah's prophecy about the virgin and Emmanuel (Is. 7, 14); Elizabeth and her son John the Baptist, the messenger of God (Lk. 1). In addition, references could be made to the themes of virginity and faith, Mary's virginity, the Church's virginity, that of the faithful, the belief that virgins can conceive, and so on—themes worked out especially by Augustine in *De Sacra Virginitate*.

When we take the scriptural themes of virginity, of barrenness, and of God as the Father of his people and of his Son, and see them in connection with original sin, in the very nature of the case they imply that the Son of God is without original sin. For where God is, there cannot be a state of non-salvation; there cannot be original sin where there is God. Thus, the entire tradition for obvious reasons says that there is no original sin in Christ.

Scripture and the Fathers of the Church appear, in this way, to lead us along a path which we too often fail to notice, or perhaps they lead us along a path which we do know toward an implication which we no longer seem able to appreciate, even though this implication is present, time and again, in Vatican Council II's *Con-*

[35] Note particularly Ps. 2, 7; Acts 13, 33; Heb. 1, 5; 5, 5; cf. also Ps. 109, 3 in the Vulgate.

stitution on the Church. To put it briefly, in the Church's tradition there is a certain obvious interchangeability between Mary the virgin mother of God's Son, the Church, virginal mother of God's people, the divine Wisdom whose joy it is to be among the children of men, and the Bride of the Song of Solomon.

If, standing on this firm ground of tradition, we are allowed to bring the same obvious interchangeability to bear on the dogma of Mary's immaculate conception (her being preserved from the *macula originalis* of hereditary sin), we see that we are declaring exactly the same thing about the Church (Eph. 5, 27), about Wisdom (Wis. 7, 26) and about the Bride (Song 4, 7); all are declared to be *"sine macula"* ("without stain"). And once again one is inclined to say, as one says of the People of God and of God's Son, that when we speak of the Church we speak of God, of community with God, or the community assembled by God, and where God is, there can clearly be no original sin, but only the stainlessness of which scripture speaks. The light of God's presence dispels all the darkness of his absence.

In our present context we cannot give more than this brief and basic summary. It is almost superfluous to mention, apart from the *"nexus mysteriorum inter se et cum fine hominis ultimo,"* the role played in this connection by that primordial element of tradition according to which scripture bears witness and proclaims not (profane) history but the truth and reality of God become man. It is only in this way that we can penetrate the patristic and medieval tradition which very often knew no Hebrew and Aramaic (not to mention other languages), which made no archeological expeditions to biblical lands, but which has set out to teach us everything about the sense of scripture and about allegories, metaphors, and demetaphorization.

Other aspects of the traditional teaching on original sin require a moment's attention. In particular, we are thinking of the so-called consequences of original sin.

The Genesis text lists the consequences of sin for the woman and for the man. For the woman they are: increased burdens of pregnancy; the pains of childbirth; domination by the man. For the man they are: the earth will bring forth thistles and thorns; he will have to earn his bread in the sweat of his brow; he will return to dust.

Tradition has always held that concupiscence, or the desires of the flesh, is the result and the sign of original sin. Where this was understood in an ethical sense, all the "passions" in general, and sexuality in particular, were taken to be visible consequences of original sin, or *"fomes peccati."*

An interesting detail is contained in this last point, and, along with it, Thomas' interpretation helps us to understand these "consequences of original sin." I refer to the autonomy of man's sexual organs, that is, the fact that they are not directly subject to the control of man's will. Augustine regards this autonomy as a result of original sin. Thomas also mentions the matter. And here we have an opportunity to state in all its acuteness the problem with which we are now dealing. Thomas not only agrees with Augustine in calling the autonomy of the sexual organs a result of original sin, but he quotes and also agrees with Aristotle's observation that this autonomy is entirely natural; he adds that one can note the same thing regarding the heart, for example, that also beats at its own pace, independently of the will's influence.[36]

Fuchs, who published a study of the sexual ethics as taught by St. Thomas,[37] remarks that the autonomy of various human organs comes from the fact that they are regulated by the so-called autonomous nervous system. This autonomy, says Fuchs, is perfectly natural and thus not a result of original sin. Fuchs, therefore, has serious criticisms to level against Thomas' position on sexual ethics. Thomas cannot make up his mind, says Fuchs, because he agrees, on the one hand, with Augustine in calling the autonomy of the sexual organs a result of original sin, and with Aristotle, on the other hand, in saying that the autonomy is natural.

When Thomas is adversely criticized, the critic himself as a rule becomes the victim of his own biased judgment, and not Thomas. This happens also with Fuchs who is quite wrong about Thomas' sexual ethic as well as about his view of original sin. Fuchs simply fails to see that, for Thomas, the terms "natural" and "result of original sin" do not imply any contradiction. For original sin is not an innerworldly reality (as Fuchs presupposes) but the *interpretation* of earthly reality in the light of revelation. This means, fundamentally and primarily, that original sin is not a particular *part* of human reality, but human reality in its entirety,

[36] *Summa Theologiae* I–II, 17, 9 ad 3.

[37] J. Fuchs, S.J., *Die Sexualethik des heiligen Thomas von Aquin* (Cologne, 1949).

precisely as excluded from the presence of God, as we have seen above.

Moreover, given the light that salvation casts on the human situation, there is nothing to prevent one from interpreting and understanding in a religious sense as non-salvation, or as "residue and consequence of original sin," all those things which stand in the way of fulfilling the meaning of human life or which stand in the way of happiness or of humanity. But this also implies precisely that men will characterize as results of original sin all those things that *appear* to them to be evil and that they *experience* as evil. Of course, these things are not always the same for all, even though there will be points on which all agree fundamentally and constantly.

It is primarily in this sense that sin is said to be the result of original sin or that original sin is said to be the cause of sin. Death is a result of original sin; so, too, are sickness, suffering, and calamities. Concupiscence, or the desires of the flesh in an explicitly sexual sense, is called a consequence of original sin, because when corporeity and "passion" are interpreted in a religious way in the light of salvation, they may be said to impede the realization of salvation, although, in themselves, they are "natural" human reality. This is why the Council of Trent did not want to speak simply of "sin," as the Reformation did, but rather chose to speak of the *"fomes peccati"* ("hearth of sin"), of *"ex peccato est"* ("consequence of original sin"), and of *"ad peccatum inclinat"* ("leads to sin"). Furthermore, Trent spoke even in this way only after it had explicitly acknowledged corporeity as a positive task for man, thus implying, in a certain sense, that corporeity has nothing to do with sin and the tendency toward evil.

In all this we find the reason why many could rightly see, in the autonomy of the sexual organs, an impediment to salvation and thus a result and sign of original sin. For it was precisely in this way that the autonomy was experienced by Augustine and by many of his own time and later (possibly on Augustine's authority). The same can be said of the consequences of the Fall listed in Genesis: the burdens of pregnancy; the pains of childbirth; the inferior position of woman; the infertility of the earth; the difficulty of daily work.

However, when material circumstances change, when the human experience of sexuality differs, and when the development of science and technology humanizes the world, then views and circumstances also change, old forms of inhumanity disappear, and new ones appear. From this it follows clearly that the consequences of original sin are or can be to some extent different now from what they were fifteen, ten, or five centuries ago.

A typical example of this is what is known as "painless childbirth." About ten years ago this appeared to be an acute problem in theology, and on January 8, 1956, Pius XII devoted one of his many addresses to the subject. Over the years a whole library grew up around the question, and in all of it the central problem derives from the concept of original sin. Some argued that painless childbirth was immoral: woman, after all, had to bring children into the world through pain. It says so in Genesis, they noted, and that is the way it must be. But the problem did not occupy thinkers for long. As usual, a practical solution was found on the basis of matter-of-fact considerations. Still, in that one problem alone, the whole concept of original sin, to say nothing of the inerrancy of scripture, was actually under review.

Thus, an investigation of the so-called consequences of original sin establishes once again the central truth that came to the fore each time we looked at a different aspect and facet of the dogma of original sin in this discussion of its main outlines. The truth is that hereditary sin (*"peccatum originale"*) is simply a synonym for the innumerable descriptive expressions in scripture to the effect that human life is meaningless as long as it is lived in God's absence, far from the promised land, far from the holy city, far from the face of God. In a sense, the absence of God is the key to all preaching and teaching on original sin, whatever form it takes.

Thus, original sin is, as it were, diametrically opposed to *virtus infusa*. Original sin is a *vitium naturae*: it is the impotence of a dead man to come to life again. For he cannot, unless the God who has become man stretches out his hand to him and commands him, "Arise!"

In the Church, the dogma of original sin is a living memorial to the bishop, pastor, and theologian Augustine. It is a monument that reminds us of one of the most important facts in the history of the Church—the victory over Pelagianism, over the naive optimism that man, through pure euphoria and the feeling of good fellowship, had become insensitive to the misery of God's absence. As a monument (in Augustine's words, *"quod monet mentem"*), it is not only a reminder of the past, but a permanent and constant confrontation of man with his divine destiny and with his personal contribution to that destiny. More tellingly and better than in any theological formulation, it is made visible for us in the immortal freshness of the Genesis story of paradise and the Fall.

Discussions on the original Reformers' doctrine of *"simul iustus et peccator"* are becoming more and more frequent, and perhaps no other context more clearly shows the constant actuality of the dogma of original sin. Since 1957, when Hans Küng published his dissertation on Karl Barth's teaching on justification, comparing it with that of the Council of Trent and suggesting, among other things, the possibility of a Catholic interpretation of *"simul iustus et peccator,"* Catholic authors have produced many similar reflections with ecumenical intent. In principle, this is a praiseworthy thing—praiseworthy as is every attempt at closer contact between Reformed and Catholic Christians. Although this contact, like all contacts among men, is not merely or even primarily a matter of theory and theology, accuracy and care in theological matters are nevertheless a prime requisite.

On both sides, however, a good deal of disastrous confusion on the question of *"simul iustus et peccator"* sometimes exists. Publications on this topic have exceptionally valuable and realistic things to say about the sinfulness of man and about justification as pure grace. Without doubt, this helps to soften many unnecessary, unrealistic, and exaggerated contrasts between Reformed and Catholic Christian teachings, or at least to reduce them to reasonable proportions.

We have seen above that good and evil are not mathematically delimited categories either in acts or in habits. Indeed, deficiencies are always to be noticed, at very least *"les défauts de la qualité,"* and these give occasion for speaking not only of good, but also of sin, and for seeing in many not only the just, but also the sinner. From the point of view of the Catholic tradition, there is, in this respect, no objection whatever to holding that man, in a certain sense, is *"simul iustus et peccator."* On the contrary, there is every reason to do so.

But one ought to remember that in this way one is expressing, albeit in a religious interpretation, a fundamentally *moral* judgment about man which, moreover, implies a kind of statistical average. It is a moral judgment that could also be formulated as follows: in spite of their faults and deficiencies, men in general are, after all, really good-willed. *"Simul iustus et peccator"* uncovers the relation to God directly implied in this humanity with all its faults, and this relation can only be of the same sort as above: it is mainly good, despite shortcomings and deficiencies.

When one uses the terms *"iustus"* and *"peccator"* in the sense of the dogma of original sin, something entirely different emerges. For then man is called "sinner" or "just" according to his existential position in the shadow and obscurity of God's absence or in the light of God's presence, quite irrespective, in the first instance, of man's *moral* level.

Considerable vagueness often arises in the use of the expression *"simul iustus et peccator."* This derives from the failure to distinguish between the moral sense of the terms "just man" and "sinner" and their strictly theological sense, or from the arbitrary confusion of ethical and strictly theological expressions. It seems to me that this indicates a lack of perception: the seriousness and the implications of the dogma of original sin are being underestimated. Augustine, were he present, would once again have serious objections against this.

3. *Mortal Sin and Venial Sin*

We must say a few words about this frequently used division. Schoonenberg suggested that the division be extended and that a third factor be distinguished: "sin unto death," of which John speaks in his first letter[38] (1 Jn. 5, 1. 17). This would indeed imply spiritual death, whereas mortal sin, although serious enough, does not imply death in the same definitive degree. Consequently, venial sin would be an insignificantly small matter.

In the light of the theological tradition, this suggestion is particularly interesting, especially if we consider how the concept of mortal sin has been devaluated in practice at the expense of the appreciation of venial sin. The traditional manuals have applied

[38] Translator's note: John's term, in Greek, is *"harmartia pros thanaton."* The Revised Standard Version of the bible translates this simply as "mortal sin," but it is not to be confused with the traditional category of mortal sin in practical moral theology.

the category of mortal sin to so many instances in so many ways (one might almost say, in season and out of season) that not only must mortal sin itself lose its force, but also hardly any room or meaning is left for venial sin. In this situation, the introduction of a "sin unto death" means the restoration of the balance of the earlier theological tradition, although it also gives rise to a problem. One cannot any longer attach much meaning to the distinction between mortal and venial sin. In other words, I have the impression that medieval theology included a great deal more in the category of "venial sin" than a later reduction to a triviality is able to realize. For the rest, only an extensive study of tradition will be able to provide any further light on this matter.

We have been considering (1) vice, human deficiency, and weakness; (2) sin, original sin, and *"simul iustus et peccator"*; (3) mortal sin and venial sin. If we try to summarize all that has been said, and at the same time remember that we are speaking of man as the origin of his own action, we arrive, briefly and concisely, at the following.

In man, vice is the lack (privation) of virtue, the lack of directedness toward his fellowmen and toward community, and this is a deficiency that in the very nature of things will make itself felt in a man's actual actions and in his contacts with other men. Behind this deficiency lies the man himself as deficient with regard to his fellowmen as well as with regard to God. Whatever qualities he may possess, man always knows that he has shortcomings as well and that he has every reason to call himself a sinner.

And yet, whatever man is and however perfect and human he may be, he is nothing, he is blind, deaf, dumb, and dead *in comparison with* that which it has never entered into human heart to conceive: light and life in God's presence. In this regard, man is not only weak but totally impotent. Here he can but receive, in the greatest wonderment, scarcely believing his own eyes.

And now all his human weakness and all human imperfection and misery are suddenly seen in a totally new light: wherever man falls short, the human face of God is obscured. This very seldom happens because of an irreparably bad will, and therefore is pardonable. But the fact remains that salvation is given into human hands, weak hands, hands that need to plead for encouragement and forgiveness again and again.

II

God as the Origin of Man's Action

A. THE OLD AND THE NEW LAW

The last part of fundamental Christian ethics deals with law and grace. At first sight it might appear strange that law, which, after all, is a human reality, should be discussed in this last part in which we are principally concerned with God as the origin of human action. One may be inclined to say either that law should have been discussed earlier or that reflection on God as the origin of human action should begin after this section on law, when we come to grace.

Such thoughts are not entirely unfounded, but fundamentally they derive from a failure to recognize the nature of this part of Christian ethics. It is true that natural law, positive law, and similar matters are often discussed here at some length, as one would expect in a treatise on law. It is also true that the treatise on law usually concentrates on interpreting the data mentioned. However, the basic inspiration of the treatise on law in this context, and particularly in Thomas' work, is an entirely different one. The subject under discussion is the divine law, the Old and the New Law, or, respectively, law and grace in the sense of John 1, 17: "For the law was given through Moses; grace and truth came through Jesus Christ." It is in this sense that we should understand the title describing the contents of this section, "Law and Grace," for the subject under discussion is indeed God as the origin of human action: *"Deus qui et nos instruit per legem, et iuvat per gratiam"* ("God who both teaches us through the law and aids us through grace").[39]

It ought not to surprise us that law, for historical and practical reasons, is rather extensively discussed in this context. Also, granting that the divine and the human are not competitive realities, it will be noted that, in a certain sense, no new ethical elements are introduced, at least insofar as it is concerned precisely with God as the origin of human action.

It is remarkable that the problem of law and grace formed a special subject in the theology of the Reformation under the title

[39] *Summa Theologiae* I–II, 90, prologue.

of "Law and Gospel," and especially *"Gesetz und Evangelium."* In Catholic theology, on the other hand, this theme is almost wholly absent, at least in the overall scheme. One either finds two successive treatises, one on law and one on grace, or—and this is more often the case—one finds a treatise on law, natural law, divine law, dispensation, *epikeia,* etc., in moral theology and a treatise on grace in dogmatic theology. Oftentimes this treatise on grace is overloaded with theological disputes about justification, merit, and similar topics, some dating from before the Reformation. This last separation of law into the field of morals and grace into the field of dogma especially caused the original scriptural theme and question of Old and New Law, law and Gospel, and law and grace to fade entirely from view.

Basically, what is under discussion? Once again, the answer is, ultimately, the mystery of God's presence in humanity. In this particular instance, it is the mystery of God as the origin of man's acts—although in a certain sense the question comes to the fore most clearly in connection with the New Testament objections, particularly Paul's, raised against the law, and also in connection with the contrast which the New Testament draws between the law and grace.

"Do you not know," Paul writes in Romans 7, "that the law is binding on a person only during his life? Thus a married woman is bound by law to her husband as long as he lives; but if her husband dies she is discharged from the law concerning her husband. . . . Likewise . . . you have died to the law through the body of Christ, so that you may belong to another, to him who has been raised from the dead in order that we may bear fruit for God. While we were living in the flesh, our sinful passions, aroused by the law, were at work in our members to bear fruit for death. But now we are discharged from the law, dead to that which held us captive, so that we serve not under the old written code but in the new life of the Spirit. What then shall we say? That the law is sin? By no means! Yet, if it had not been for the law, I should not have known sin. . . . Apart from the law, sin lies dead. . . . The very commandment which promised life, proved death to me. . . . The law is holy, and the commandment is holy and just and good. . . . We know that the law is spiritual. . . ."

Romans 8 continues: "There is therefore now no condemnation for those who are in Christ Jesus. For the law of the Spirit of life in Christ Jesus set me free from the law of sin and death. For God has done what the law, weakened by the flesh, could not do; sending his own Son in the likeness of sinful flesh and for sin, he condemned sin in the flesh, in order that the just requirements of the law might be fulfilled in us, who walk not according to the flesh but according to the Spirit."[40]

The law is powerless; it only makes sin known. And yet the law is holy and even spiritual. It is given by God.[41] Nevertheless, "no human being will be justified in his [God's] sight by works of the law."[42]

What does Paul mean? For, indeed, he has written in the same letter: "For it is not the hearers of the law who are righteous before God, but the doers of the law who will be justified" (Rom. 2, 13). Is one then justified after all by observance of the law? A moment ago we heard him maintaining the opposite.

Christ's word is well known: "Think not that I have come to abolish the law and the prophets; I have not come to abolish them but to fulfill them. For truly, I say to you, till heaven and earth pass away, not an iota, not a dot, will pass from the law until all is accomplished" (Mt. 5, 17. 18). The Letter to the Hebrews, however, appears to be more radical: "For when there is a change in the priesthood, there is necessarily a change in the law as well" (Heb. 7, 12); ". . . a former commandment is set aside because of its weakness and uselessness (for the law made nothing perfect)" (Heb. 7, 18. 19); "In speaking of a new covenant he [Jeremiah] treats the first as obsolete. And what is becoming obsolete and growing old is ready to vanish away" (Heb. 8, 13).

That "new covenant" is described by Jeremiah in this way: "This is the [new] covenant which I will make with the house of Israel after those days, says the Lord: I will put my law within them, and I will write it upon their hearts; and I will be their God, and they shall be my people. . . . They shall all know me, from the least to the greatest, says the Lord; for I will forgive their iniquity, and I will remember their sin no more."[43] This reminds us of Paul's word to the Corinthians: "You yourselves are our

[40] Rom. 7, 1. 2. 4–7. 8. 10. 12. 14; 8, 1–4.

[41] Cf. Mt. 15, 6; Lk. 24, 44; Jn. 5, 46; St. Thomas, *Summa Theologiae* I–II, 98, 2.

[42] Rom. 3, 20; cf. Ps. 143, 2; Gal. 2, 16.

[43] Jer. 31, 33. 34; cf. Heb. 8, 8–12.

letter . . . a letter from Christ delivered by us, written not with ink but with the Spirit of the living God, not on tablets of stone but on tablets of human hearts" (2 Cor. 3, 2. 3).

When the New Testament authors speak in this or any other way of the Old and the New Law, the Old and the New Covenant, law and grace, law and Spirit, letter and Spirit, flesh and Spirit, law and faith, or sin and justification, they do it, to be sure, in terms of the "scriptures"—that is, of the Old Testament—but they are very well aware that they are proclaiming its fulfillment. They bring something new, something for which previous generations had longed, which previous generations had announced and foretold, but which now, for the first time, is revealed and made known: ". . . the mystery of Christ, which was not made known to the sons of men in other generations, as it has now been revealed to his holy apostles and prophets by the Spirit" (Eph. 3, 5; cf. Col. 1, 26).

All that is said about the law is therefore said in the light of the mystery of Christ, and it must be understood in this light.

Thus it becomes clear, first of all, that sin and justification in this context must not be interpreted primarily in a moral sense; on the contrary, they express salvation and non-salvation. In other words, the primary significance here of sin is "original sin" or the absence of God, and justice or justification is the opposite of this, namely, the presence of God. The law, not merely in the sense of commandment or precept, but as a proclamation of salvation, does not itself bring this salvation, this presence of God, or this justification. It is much rather the revelation of sin or of the absence of God. "The letter kills" (2 Cor. 3, 6) is something that must be said not only of the Old Law, but also of the New, not only of the law but also of the Gospel, as Augustine and Thomas, among others, bear witness.[44]

What is actually new in the New Testament and in the New Law is "the grace of the Holy Spirit through faith in Christ," as Thomas says, in these few words saying all that can be said.[45] For Jeremiah speaks of a law written in the heart, and Paul speaks of the law of the Spirit (Rom. 8, 2) and of the law of faith (Rom. 3, 27);

[44] *Summa Theologiae* I–II, 106, 2.

[45] *Ibid.*, I–II, 106, 1: *"gratia Spiritus Sancti quae datur per fidem Christi."*

moreover, Augustine says that God's laws, written in the heart of man, are nothing other than the presence of the Holy Spirit.[46]

Was the Spirit of God therefore absent before—one might be inclined to ask—while the Old Testament "from the beginning" was proclaiming his presence? Were there then no "believers," even though Abraham stands as the protoype of the believer and even though the Letter to the Hebrews lists a whole chapter of believers (Heb. 11), a "cloud of witnesses" (Heb. 12, 1)?

The answer clearly depends on the view that one holds of the difference between the Old and the New Law, between the Old Testament and the New. We are accustomed to think of a chronological difference, with the Old Testament passing and the New emerging more or less around the year 1 by the current method of numbering years. But if this is the distinction we make, the proclamation of salvation becomes unintelligible, because it is replaced by mere history.

The *"unitas fidei utriusque testamenti"* ("the identical faith in both testaments")[47] belongs strictly to the conviction and faith of the entire tradition of the Church, and in the 12th century it formed, among other points, an important theme for theological reflection. In this connection, Thomas made a number of very striking statements in which he regards every believer as belonging, in a certain sense, to the New Testament.[48] Neither does he have any objection to the converse, in a certain sense, which is that unbelievers in the New Testament belong to the Old.[49] In the same

[46] *De Spiritu et Littera,* 21; see St. Thomas, *loc. cit.,* I–II, 106, 1.

[47] St. Thomas, *Summa Theologiae* I–II, 107, 1 ad 1.

[48] "*. . . nullus unquam habuit gratiam Spiritus Sancti nisi per fidem Christi explicitam vel implicitam. Per fidem autem Christi* pertinet *homo* ad novum testamentum. *Unde quibuscumque fuit lex gratiae indita, secundum hoc* ad novum testamentum pertinebant" (*Summa Theologiae* I–II, 106, 1 ad 3, repeated in 106, 3 ad 2); *"Illi autem qui in veteri testamento Deo fuerunt accepti per fidem, secundum hoc* ad novum testamentum pertinebant; *non enim iustificabantur nisi per fidem Christi, qui est auctor novi testamenti"* (*ibid.,* I–II, 107, 1 ad 3).

[49] *"Fuerunt tamen aliqui in statu veteris testamenti habentes caritatem et gratiam Spiritus Sancti, qui principaliter exspectabant promissiones spirituales et aeternas. Et secundum hoc* pertinebant ad legem novam. *Similiter etiam in novo testamento sunt aliqui carnales* nondum pertingentes ad perfectionem novae legis, *quos oportuit etiam in novo testamento induci ad virtutis opera per timorem poenarum, et*

context he makes a very pointed remark concerning the mystery of Christ.[50]

The range and the intent of the problem under present consideration begins gradually to take on clearer form and contour. God's saving work and his epiphany in this world are not confined to particular times and places. Thus Thomas, in speaking of redemption, will refer to the power of God, the *"virtus divina,"* which *"praesentialiter attingit omnia loca et tempora"* ("is present in and embraces all times and places"),[51] thus using in his christology the same terms as those in which God's omnipresence is defined.[52] Scripture paints God's saving work in mankind in shades and colors to bring out and proclaim its truth and reality, and not in order to put restraints and limits to its generality, universality, and catholicity.

Man receives all salvation, grace and faith, even in the time of the law, from the God who has made himself manifest in human form, and from him only, for "there is one mediator between God and men, the man Christ Jesus" (1 Tim. 2, 5). This is the New Testament, or, in other words, it is the grace of the Holy Spirit through faith in Christ. In relation to this all-embracing saving work and saving presence of God, the proclamation and the law are secondary.

Does not the Old Testament, then, fall away, if God's work of grace, which the New Testament is, embraces the whole history of men and of the world? The answer cannot be a simple yes or no. There are not only various aspects to be distinguished, but the answer also will always, ever and again, depend on the way in which we understand the difference between the Old and the New Testaments. It is not merely a question of whether or to what extent Israel lived under the Old or under the New Law. It is a question also of whether we think that we live exclusively under the New Law. For if a chronological or geographical distinction between Old and New Law does not correspond to the reality,

---

per aliqua temporalia promissa" (*ibid.*, I–II, 107, 1 ad 2), i.e., in the Old Testament fashion.

[50] *Ibid.*, I–II, 103, 2: *"Poterat autem mens fidelium, tempore legis, per fidem coniungi Christo* incarnato *et* passo; *et ita ex fide Christi iustificabantur."*

[51] *Ibid.*, III, 56, 1 ad 3.

[52] Cf. *ibid.*, I, 8; I *Sent.* 37; Peter Lombard, I *Sent.* 37, 1.

then all simple schemes and explanations based on that distinction are wiped out.

Therefore, let us leave this presupposition entirely open, and ask the question afresh. Does the Old Testament fall away because the New Testament—that is, the grace of the Holy Spirit—embraces the whole history of man? In general, the answer must be negative: the Old Testament does not fall away. But some closer explanation is necessary.

(a) Inasmuch as we are speaking of the grace of God, we are not speaking of the Old Law or the Old Covenant, but exclusively of that which is most essential and central in the New Testament: the "law of the Spirit" (Rom. 8, 2), justification through faith in Christ. It is not only the Old Law that is irrelevant in this context, but also the New, precisely insofar as its character as law is concerned—that is, insofar as the New Law involves both precepts and proclamation. Of both it is true that the letter kills; the letter does not bring about man's justification or salvation.[53] In this respect, Old and New Law do not differ. Moreover, as precept they do not differ from any other law whatever, even though they certainly do differ from all other law, and from each other, too, precisely insofar as they are the proclamation of salvation.

Therefore, in speaking of the Old and the New Testaments, we must distinguish the following: (1) the mystery of Christ as an actual reality: salvation, faith, and grace; (2) the explicit proclamation of salvation in the New Testament; (3) the implicit proclamation of salvation in the Old Testament; (4) the precepts of the New Testament; (5) the precepts of the Old Testament.

(b) The Old Law, as the implicit proclamation of the mystery of Christ, did not and does not fall away insofar as the explicit proclamation of the mystery of Christ had (or has) not yet taken place. The validity of the implicit proclamation of salvation is not canceled out by the actual saving work of God in Christ. On the contrary, this implicit proclamation is part of the actual reality of God's saving work.

(c) Concerning the precepts of the Old Testament, Thomas and others make the traditional distinctions between "moral," "ritual," and "juridical" precepts. *"Praecepta moralia"* ("moral commandments") are practically the same as the natural law.[54]

[53] Cf. *Summa Theologiae* I–II, 100, 12 and 106, 2.
[54] Cf. *ibid.*, I–II, 100, 1.

"Praecepta caeremonialia" ("ritual prescriptions") are a determination, in greater detail, of the natural law with regard to man's relationship to God.[55] *"Praecepta iudicialia"* ("juridical codes") are a determination, in greater detail, of the natural law with regard to the relationships of men among themselves.[56]

The conclusion is obvious. The Old Law, as far as its moral precepts are concerned, does not fall away. For these precepts are natural law, and this remains in force. However, the ritual prescriptions are rendered obsolete by the proclamation of the mystery of Christ, for their purpose was to prefigure this mystery, and they are superseded by other ceremonial rites which now signify and represent the revealed mystery of Christ.

The same may be said of the juridical code as prefiguring the mystery of Christ. Nevertheless, as social norm, it may possibly retain its validity, even after the revelation of the mystery of Christ.[57] However, it is somewhat improbable that it should, both because social norms must be suited to the community and because, for this reason, they are subject to amendment and correction within the same community.[58]

Thus, the ritual prescriptions are fulfilled rather than rejected by the New Law, and to a certain extent the same may be said of the juridical precepts.

The presence everywhere and at all times of the grace of the Holy Spirit does not therefore abrogate the Old Testament; on the contrary, it gives the Old Testament its character as a preliminary indication of what is to come. Even with the revelation of the mystery of Christ, the Old Law retains its force to some extent—that is, as far as its moral precepts are concerned, for these are natural law. Besides these, the New Law contains "sacraments," but no "ritual prescriptions" or "juridical code," according to Thomas.[59]

What does this thesis imply as far as the "sacraments" and the "moral precepts" are concerned? If the relationship of the sacraments to the ritual prescriptions is the same as the relationship of natural law to positive law (community norms, or *"praecepta*

[55] Cf. *ibid.*, I–II, 101, 1.
[56] Cf. *ibid.*, I–II, 104, 1.
[57] Cf. *ibid.*, I–II, 104, 2 and 3.
[58] Cf. *ibid.*, I–II, 104, 3 ad 1 and ad 2.
[59] Cf. *ibid.*, I–II, 108, 2.

iudicialia"), then the content of the New Law is, in a certain sense, much more general and possibly also much more concrete than is often thought.

Therefore, we must now turn our attention to law if we wish to understand how God, in fact, is the concrete origin of human action. In particular, we shall be concerned with the instructive function of both natural law and positive law and with the relation of the one to the other.

The concept of *natural law* dates from the classical period of Greek philosophy. Since that time it has undergone many changes in its meaning. According to Hans Leisgang,[60] the history of natural law began with Antiphon the Sophist (between 450 and 400 B.C.), who opposed "nature" and "law" in the same way as "reality" and "appearance," intending to say by this that law destroys the natural equality of men. Plato objected to this on various grounds, noting particularly that this "return to nature" leads necessarily to the supremacy of the most powerful, while it is precisely the weak ones whom the law protects. Aristotle also rejected the opposition of "nature" and "law," although he did make a distinction between the "naturally just" and the "legally just." But these are not two separate and materially discrete realities or norms; they refer to two formally distinct ways of looking at the norms in actual force. For, on the one hand, norms in actual force represent a "natural" requirement of man as a social being; on the other hand, the precise, concrete, and variable form and content of these norms depends on the actual community itself and can be different in different communities. In other words, the norms in actual force are "natural" in a certain sense and also "legal" in another sense. Here "legal" means that which belongs to positive ordinance or law ("law" not yet having the limited meaning of a written law, in accordance with the narrowed definition the word came to have during Aristotle's lifetime.[61] Nature and law are thus not opposed in the way that "necessary" and "arbitrary" are opposed, but they are two aspects of every human norm and relate to each other as the universal to the concrete; in a certain sense also, they relate to

[60] Cf. "Physis," in Pauly-Wissowa, *Realenc. d. Klass. Alt.* XX (Stuttgart, 1941), pp. 1129–1164.

[61] Cf. A. R. Henderickx, O.P., "De algemeene rechtvaardigheid in de Nikomachische Ethiek van Aristoteles," in *Tds. v. Philos.* I (1939), pp. 277–318, esp. p. 284.

each other as that which is to be determined relates to that which determines.

One notices, however, particularly among the Sophist philosophers, that from the very beginning there was a tendency to take "natural law" as a body of law on its own, with its own particular content, alongside of and apart from the positive norms and laws of particular communities and societies. By way of the Stoics, Cicero, Roman law, and Isidor (the great anthologist of antiquity), this tendency to take natural law as a substantial and independent reality endured stubbornly in the Latin and Western tradition.

It was not only a philosophical tradition, however, that brought natural law to the West. A theological tradition, going back to Philo in the great city of Alexandria around the beginning of the 1st century, identifies the natural law with the Decalogue and by preference expresses the natural law in the so-called "golden rule" which was more ancient than its negative fomulation in Tobit 4, 16 and its positive expression in Matthew 7, 12: "Whatever you wish that men would do to you, do so to them; for this is the law and the prophets."[62] Gratian also begins the first distinction of his *Decretum* in this way.[63] The theological tradition of the West, partly due to Gratian's influence, would continue not only to speak of natural law, but also to consider natural law as a kind of distinct reality and to identify it with the "golden rule" and, more especially, with the Decalogue or the *"praecepta moralia"* of the Old and New Law.

The precision of Aristotle also appears in tradition—in Thomas among other writers. Like Aristotle, Thomas makes a formal rather than a material distinction between natural law and positive law. This shows why his formulation of the natural law in a certain sense includes everything and in another sense includes nothing: *"Bonum est faciendum . . . et malum vitandum"* ("Good must be done, evil avoided").[64] This is identical with *"secundum naturam agere"* ("act according to nature") and with *"secundum rationem agere"* ("act according to reason").[65] The concrete content of such

[62] Cf. A. Dihle, *Die Goldene Regel . . .* (Göttingen, 1962).

[63] *"Humanum genus duobus regitur, naturali videlicet iure et moribus. Ius naturae est, quod in lege et evangelio continetur. . . ."* He then cites Mt. 7, 12.

[64] *Summa Theologiae* I–II, 94, 2.

[65] Cf. *ibid.*, I–II, 19, 3, etc.

precepts is not arbitrary in the least, but depends entirely upon the concrete community and the actual human situation.

In this way, Thomas can agree with the tradition and see the whole of the Old Law summed up in the "golden rule"[66] which, in its turn, is implicit in the commandment, "You shall love your neighbor as yourself" (Lev. 19, 18). For, indeed, Paul says of this: "He who loves his neighbor has fulfilled the law. The commandments, 'You shall not commit adultery, You shall not kill, You shall not steal, You shall not covet,' and any other commandment, are summed up in this sentence, 'You shall love your neighbor as yourself' " (Rom. 13, 8. 9). In this love of neighbor is included the love of God, as Thomas declares, continuing his explanation in the same context.[67] In this respect, therefore, the New Law is identical with the Old.

The distinction, then, between the New and the Old Law is, in a certain sense, a negative one. The New Law knows only the natural law; it knows no "juridical precepts," no particular and concrete prohibitions and commands of the kind that can differ in different communities. In other words, all particularism is foreign to the New Testament. The New Testament is general and universal; it is catholic. It is directed to all men, regardless of culture, ethos, history, or anything else. But precisely because of this universality, it is able to enliven and enlighten every particular and concrete determination of the natural law with the light of God's offer of salvation and of his saving appeal to man.

The New Law, in a certain sense, does not even know the natural law or, consequently, any application or determination of the natural law, except precisely insofar as the grace of God needs to be given a human form in some way, so that it may become God-with-us and grace-for-us.[68] This is impossible without a minimum of humanity, and in itself it tends toward a maximum of humanity.

More than any other law, and more even than the Old Law, the New Law of grace calls for "virtue" or genuine humanity,[69] not (let us say it again) for the sake of authentic humanity as such, but

[66] *Ibid.*, I–II, 99, 1 ad 3.
[67] *Ibid.*, I–II, 99, 1 ad 2; cf. 100, 3 ad 1; II–II, 44, 2; 105, 2.
[68] *Ibid.*, I–II, 108, 1 and 2.
[69] Cf. *ibid.*, I–II, 90, 3 ad 2; 92; 94, 3; 95, 1; 96, 2 ad 2; 96 3c and ad 3; 107, 2 and 4.

for the sake of the faith that works in love,[70] the justification infused by God[71]—that is, faith, hope and love,[72] or the divine virtues of which we spoke above, which simply cannot exist without authentic humanity. For indeed it is God himself who turns one's acquired humanity into "righteousness" in the sense explained, and he does this through his manifestation in humanity and through the grace of his Holy Spirit.

That which holds true for natural law and community norms (*"praecepta iudicialia"*) also holds true for the ritual prescriptions (*"praecepta caeremonialia"*) which are not, however, laid down by the New Law itself, but left to men or to the competent authority. The question does not concern us immediately in this context, but one might well ask what this means, since it is commonly held that the "sacraments" are laid down and determined by the New Law. If the sacraments imply no ritual prescriptions, what do they imply? Whatever else the answer may involve, they imply, first of all, the proclamation of the mystery; but in saying this we are doing hardly more than translating the word "sacrament" back into Greek. But beyond this, what?

If we try to summarize the above in keeping with the basic topic of this discussion—God as the origin of human action—then it becomes clear that God, as creator, is the origin of autonomous man and of his autonomous humanity.

However, what we call the New Law, which God himself writes in the hearts of men, is formed only by the manifestation of God himself in the world of men and by the power of the Spirit of God who supports men in their autonomy. The relevant fact here is not the metaphysical axiom that God, as *causa prima,* or first cause, is origin of all creaturely being and action, but rather the salvational fact that God, through his Spirit, leads men into community with himself. Rightly enough, it is the same God who is the subject of both of these statements, but the total implication of God's work comes to light only in the situation of revealed salvation.

In other words, the New Law transcends all that we normally understand by law, and actually, in its totality, it is not a law. Nevertheless, if the term is used, as in New Law, or law of the Spirit, this is largely, if not entirely, due to the influence of Jeremiah

[70] Gal. 5, 6; cf. St. Thomas, *ibid.,* I–II, 108, 1.
[71] *Ibid.,* I–II, 100, 12.
[72] *Ibid.,* I–II, 103, 3.

and the imagery he employs when speaking of the law which God will write in the hearts of men: "And no longer shall each man teach his neighbor and each his brother, saying, 'Know the Lord,' for they shall all know me, from the least of them to the greatest" (Jer. 31, 34; Heb. 8, 11). God is not a stranger to man, for man finds him and acknowledges him in his own human world, although only through the power of God's Spirit who "is at work in us, both to will and to work" (Phil. 2, 13), but who is also given to us "as a guarantee" (2 Cor. 5, 5).

We have seen above that the divine and the human are not in competition, and this applies equally in the matter of the law. The human reality of law as a whole is not being threatened when we speak of the New Law. Or perhaps, we could state it better by saying that human law, and the human task which it involves, is, just like the total reality of man and world, part of the new earth that arises through God's epiphany. This can be only because God does not infringe on man's autonomy, but establishes it as an essential condition and part of his saving work.

It seems to me that this provides us with the key to the meaning of Thomas' remarkable observation that people of the Old Testament belonged to the New, and also that people in the New Testament do not yet, in a certain sense, belong to the New Testament. This can be explained in terms of the human tension between law and virtue. The purpose of law is to lead toward virtue and authentic humanity, and where virtue is perfected, law in a certain sense has no further task or function and is no longer necessary. Man himself has then become law, as Paul writes in his Letter to the Romans (2, 14).[73]

Now, if something is true of human law and virtue, it must also be true of the Old Law and the New Law, or, rather, of the Old Law and grace, or faith, or justification. For where there is virtue or authentic humanity, there is the grace of the Holy Spirit and justification, as the revelation of salvation makes clear. Where Christ comes to live, the "pedagogue," the "custodian," is no longer needed.[74] The law has the community of men in view, and revelation shows that this community is community with God, and in community with God the law has no further function. "If you

[73] Cf. Rom, 2, 15; 8, 14; Gal. 5, 18; St. Thomas, *Summa Theologiae* I–II, 90, 3 ad 1; 93, 6 ad 1; 96, 5c and ad 1, ad 2.

[74] Cf. Gal. 3, 24; 2, 20; 4, 19; Col. 3, 4; etc.

are led by the Spirit," writes Paul, "you are not under the law" (Gal. 5, 18). "The fruit of the Spirit is love, joy, peace, patience, kindness, goodness, faithfulness, gentleness, self-control; against such there is no law" (Gal. 5, 22. 23).

Where authentic humanity is still lacking or where it falls short, however, there is still a function for the Old Law, and there is still a lack of the perfection of the New Law of the Spirit[75] which, as a matter of fact, cannot be more perfect in a given individual than the humanity in which that perfection takes form.

Thus, the contours of this treatment of Old and New Law gradually become more sharply and clearly defined the more we relate the numerous texts, formulations, and interpretations back to the human principle of tension between law and virtue. It is this principle that, in theological and salvational interpretation, covers and contains the whole theme of Old and New Law, law and grace, and letter and Spirit.

It should be clear that there is no possibility of drawing any naive or schematic chronological borderline between Old and New Law. If something is simply not true in human reality itself, it cannot suddenly be made true by any imposition of "supernaturalistic" theory. At the same time, it should also have become clear, at least to some extent, how we are to understand the urgent appeal in Protestant theology, where it speaks of the dialectic between law and Gospel.

It must be affirmed beyond doubt that God himself, and not man only, is the origin of human action, and indeed that its primary origin is God. But, at the same time, it is clear that there is still a long way to go to the ultimate goal that Paul describes as that total subjection which will make God "everything to everyone" (1 Cor. 15, 28).

B. GRACE

In terms of content, the treatise on grace is actually of less importance than the treatise on the Old and New Law, however remarkable such a statement may seem at first sight. Perhaps, however, now that we have completed the previous section, the statement may not seem so remarkable. In the main, the treatise on grace contains only a number of elements taken from the treatise

[75] Cf. St. Thomas, *Summa Theologiae* I–II, 107, 1 ad 2.

on the Old and New Law and determines them more precisely and in greater detail, as required, at least to some extent, by particular historical problems and formulations. It may be that the description which Thomas gives of the subject matter of this whole last section is "God who instructs us through the law and aids us through grace," but from the actual content of the treatise on the Old and New Law, it is already clear that God's instructing us through the law is ultimately the same thing as God aiding us through grace.

Two points seem important enough to require our special attention here: (1) the description of grace as accident; (2) merit.

1. *The Description of Grace as Accident*

This description, as far as I know, appeared for the first time in medieval theology, when various philosophical and, particularly, Aristotelian categories began to make their influence felt. Grace then became variously described as a *"qualitas,"* as a *"forma accidentalis,"* and as *"accidens."* Subsequent theology adopted these terms, and as late as 1963 one finds C. Baumgartner writing, in what purports to be an interpretation of Thomas on the subject of grace, *"c'est une forme accidentelle, une* 'qualité' *surnaturelle."*[76] And it is certainly true that Thomas also uses these terms. He calls grace a *"qualitas,"*[77] a *"forma accidentalis,"*[78] and an *"accidens."*[79]

Before one draws any absolute conclusions, however, from the use of the terms themselves, divorced from the context in which they occur, the precise sense in which the terms are used, and particularly by Thomas, should be clarified. Popular interpretation would have grace understood as an accident in the substance man, and thus as a non-essential extra added to man already in existence. Quite a number of years ago, this view already had its critics.

How does Thomas use these terms? When he adopts the term *"forma accidentalis,"* he does not say that a non-essential *"forma,"* grace, is added to man. He does say: "That which is, substantially, in God, comes to be, accidentally, in the soul which participates in

[76] *La grâce du Christ* (= Le Mystère Chrétien. Théol. dogm. 10) (Tournai, 1963), p. 91, n. 4, referring to *Summa Theologiae* I–II, 110, 2 and 110, 3 ad 3.

[77] *Ibid.*, I–II, 110, 2c and ad 1; 110, 3 ad 3.

[78] *Ibid.*, I–II, 110, 2 ad 2.

[79] *Ibid.*, I–II, 110, 2 ad 3.

the divine goodness" (*id . . . quod substantialiter est in Deo, accidentaliter fit in anima participante divinam bonitatem*).[80] In other words, "accidental" is not opposed to "substantial" as referring to man, as the popular interpretation would have it, but to "substantial" as referring to God. Man receives a share in the nature of God, according to 2 Peter 1, 4. The text is very familiar in this context,[81] and it is cited once again by Thomas. It is precisely in the sense of this text that reason can be found for speaking of grace as accidental with regard to man. Thomas uses the concepts of "*accidens*" and "*qualitas*" in the same manner.

In other words, when grace is spoken of in this way, there is no question of something incidental and non-essential being added to man who remains unaltered in other respects. The original meaning of the terms is in thorough agreement with Paul when he says, "If anyone is in Christ, he is a new creation";[82] he is no longer a "man of dust," of the earth, but a "man of heaven" (cf. 1 Cor. 15, 47. 48); "once . . . darkness, now . . . light in the Lord" (Eph. 5, 8); "delivered . . . from the dominion of darkness and transferred . . . to the kingdom of his beloved Son" (Col. 1, 13).

The whole of man is clearly grace. If one still wants to use the term "accidental," the only thing that one can mean is that which distinguishes man from God. It does not refer to any particular and quite unintelligible sector of man as distinct from other, more real, sectors.

The qualification of grace as an accident by a large number of authors must stand as a classical example of superficial and unhistorical interpretation, not only for the reasons already given, but also and above all for reasons that will become clear as we examine the historical situation of the problem of grace in medieval theology. For this bears the marked imprint of Peter Lombard's identification of love with the Holy Spirit.[83] This gave emphasis to the questions of whether grace is, in fact, something *in* man, whether it does anything *in* man, and whether it is something permanent *in* man. The terms in the original context thus have an

[80] *Ibid.,* I–II, 110, 2 ad 2.

[81] *Ibid.,* I–II, 110, 3.

[82] 2 Cor. 5, 17; cf. Gal. 6, 15; see also St. Thomas, *ibid.,* I–II, 110, 2 ad 3; 112, 2 ad 3.

[83] Cf. J. Auer, *Die Entwicklung der Gnadenlehre in der Hochscholastik* . . . I (Frieburg, 1942), pp. 109ff.

entirely different emphasis from that placed on them in a later context. There the established principle was the effect in man, and the question of *how* the effect is brought about was unfortunately and much too one-sidedly dominated by terms like "accident" which had become divorced from their proper connections.

2. *Merit*

The Latin term *"meritum"* played a very important role in Western theology, particularly at the time of the Reformation and later. The term does not come from scripture (in this case, the Vulgate) where the verb *"mereri"* occurs very infrequently. The term came into the theological tradition through Tertullian who used it, however, in a sense that was still neutral and applicable to evil as well as to good, more or less in the way in which we use the word "deserve."

In Cyprian, the term *"meritum"* has something of its later, somewhat technical meaning, but unquestionably this "merit" is entirely a matter of God's grace and benevolence.[84] Augustine uses the term very often and always with the same precise meaning as found in Cyprian and, indeed, in all the Fathers: merit is a divine gift.[85] In Chapter 16 of the famous *Decree on Justification,* the Council of Trent takes over from Augustine the formulation that merits are gifts of God.[86] Canon 18 of the Second Council of Orange in 529 is equally clear.[87]

It was only in the 12th century that theologians began to discuss merit in a more detailed and less traditional way and to make distinctions between *"meritum de condigno," "meritum de congruo," "meritum interpretativum,"* and so on. In stating this, we must re-

[84] "*. . . remunerans in nobis quicquid ipse praestitit et honorans quod ipse perfecit*" (Ep. LXXVI, 4: ed. Hartel, CSEL III/2, 831).

[85] Augustine, for example, writes: *"Diffundebatur in eis caritas per Spiritum Sanctum* (Rom. 5, 5), *qui ubi vult spirat* (Jn. 3, 8), *non merita sequens, sed etiam ipsa merita faciens*" (*De gratia Christi et de peccato orig.* II, 24, 28: PL 44, 399).

[86] "*. . . tanta est erga omnes homines bonitas, ut eorum velit esse merita, quae sunt ipsius dona*" (Denz. 1584).

[87] *"Debetur merces bonis operibus, si fiant: sed gratia, quae non debetur, praecedit, ut fiant"* (Denz. 388); taken over from Prosper of Aquitane, *Sententiae ex Augustino delibatae,* 297: PL 45, 1885—practically a literal repetition of St. Augustine, *Contra Jul. opus imperf.* I, 133: PL 45, 1133.

member that semi-Pelagianism (and also the text of the Second Council of Orange of 529) was unknown to medieval theologians until Thomas came across it in one of his many historical researches. According to Bouillard, this was probably around 1260, during a stay in Italy.[88] The text of Orange was apparently unknown until Crabbe produced his edition of the Councils in 1538, of which the Fathers of the Council of Trent made immediate use.

The fact remains that Thomas was certainly much more sober in his explanation of *"meritum"* than many of the medieval theologians. For instance, he distinguishes between *"meritum de condigno"* and *"meritum de congruo,"* but he says hardly more of *"meritum de congruo"* than that it is not "merit," and in this Thomas stands in contrast to other theologians who tended to develop and expand the concept of "merit" in all sorts of ways. If later theologians had followed Thomas' line instead of the line represented by the 12th-century division of *"meritum de condigno"* and *"meritum de congruo"*—a line maintained up to our own day—the theological controversy between Rome and the Reformation on this point would, in all probability, have been much less acrimonious and of much shorter duration.

The discussions on merit can conceal a certain exaggerated one-sidedness, for they are often based on a biased interpretation of the term "merit" as it is found in the theological tradition. When tradition speaks of "merit," very often the meaning in the first instance has nothing to do with "deserving heaven" or the "accumulation of merits," expressions which not very long ago were casually bandied about. The term *"meritum"* is very often the obverse of *"peccatum"* (sin)—that is, a good human act is called *"meritum"* in relation to God, whereas a bad human act is called *"peccatum"* in relation to God. When *"meritum"* is used in this way, it is not exactly the equivalent of the English word "merit," which has come to have too strong and particular a meaning.

Such phrases as "the accumulation of merit" belong to a figurative manner of speaking in which the actual content and significance of merit are not made entirely clear. To effect this result, a penetrating investigation through to the personalist sphere of human action is necessary. Then "to deserve" and *"mereri"* receive again their original meaning of "to serve" and "to make oneself

[88] *Conversion et grâce chez S. Thomas d'Aquin* (Paris, 1944), pp. 103–108.

useful," and this is not, in the first instance, something that calls for an extrinsic reward, but something that gives rise to an intrinsic relation and direction toward other people and toward community, and thus toward God. Any possible extrinsic effect of such action is, of itself, a secondary matter. One can say that *"meritum"* (merit) is essentially nothing more than an improving of personal relationships. This is why Thomas can say of merit, *"Reducitur ad dispositionem materiae"* ("It can be compared to a disposition of matter").[89] Where God is concerned, this improving of personal relationships is always embodied in relations with fellowmen in humanity shared in common. This needs no further confirmation.

Human growth takes place in the service of God, of fellowmen, and of community. This is the anthropological basis upon which one was able to use phrases like "the gaining and accumulation of merit," although the phrases have a distastefully commercial flavor in the context. It is hardly necessary to say explicitly that this growth, not only as a human reality, but also in its relation to God, is supported entirely by the grace of God.

The endurance of this relation to God after a man's death cannot be said to be merited in any sense, except in terms of the pure gratuity of God's plan of salvation worked out in us by the power of his Spirit "who is the guarantee of our inheritance."[90] Thus there is no justification whatever for speaking of "merit" in any but the most modest of terms.

This discussion of law and grace helps us to see the relation of the Old Law to the New Law in terms of the human model of the relation of law to virtue. In principle, law becomes superfluous only where virtue has come to perfection—and therefore law is, in fact, still necessary and will always remain necessary and useful. In the same way, the Old Law is, in principle, superfluous only where the grace of the Holy Spirit rules the entire man—and so the Old Law, insofar as it is in a certain sense perpetuated in the New Law, is, in fact, necessary and useful as a custodian leading to Christ (Gal. 3, 24), "until we all attain to the unity of the faith and of the knowledge of the Son of God, to mature manhood, to the measure of the stature of the fullness of Christ" (Eph. 4, 13).

God is at the origin of human action, particularly through the practical "instruction" of his actual engracing and saving work.

[89] *Summa Theologiae,* I, 23, 5.
[90] Eph. 1, 14; cf. St. Thomas, *ibid.,* I–II, 114, 3c and ad 3.

Thus, with his own hands he forms the image and likeness of God in endless variety. This is God's plan, and it is paradise, even if we can see no farther than its outer edges and even if these edges do not often seem to be much like paradise.

It belongs to the function of a fundamental Christian ethic to see the outer edges, the human side, as it really is, with all its human factors and implications. Only in this way will the Christian ethic be able to fulfill its most essential task, to be a theology, to be a reflection on the God who hides himself from the eyes of those who look up to the clouds, but who wants himself to be found in the work of his own hands, the work he himself found "very good," and who desires that result not for any special reason, but just because he wants it so. Actually, our complicated world has no meaning apart from this.

That this should lead us to practical conclusions is surely not excluded. It is hardly likely, however, that we would ever discover God in his works if we were to begin by hammering our heads against the stone wall of "moral problems."

PART III
Special Christian Ethics

1. Divine and Cardinal Virtues

If one disregards the treatise on the sacraments, which for very practical reasons is given a place in the manuals of moral theology (sometimes filling an entire volume), the three theological virtues and the four cardinal virtues are the main points of the outline in which the treatise on so-called special ethics or morals is constructed. The point and aim of our inquiry here is to look into the fundamental construction of this treatise; it is simply not possible to go into every detail of special ethics.

This presents us with four questions. Their concern will be: (1) the relation of the divine virtues to the cardinal virtues; (2) the three divine virtues; (3) the four cardinal virtues; (4) conscience.

I

The Relation of the Divine Virtues to the Cardinal Virtues

Fundamental Christian ethics is entirely dominated and determined by the Christian reality, or, more accurately, by the Christian character of the human reality. What appears to be merely human communication and community, human action and intersubjectivity, genuine humanity and human deficiency—in brief, pure humanity—turns out to be, in reality, community with God and Church, acknowledgment and rejection of God himself, life or death—in brief, total involvement in a world and a universe which is the form and expression of God's almost extravagant care and generosity, and life in personal intimacy with him.

Special Christian ethics turns its attention, first of all, to this last fact and begins with a treatise on the divine virtues. Next, it considers the concrete, earthly implications and form of this "spiritual" life in the treatise on the cardinal virtues. Therefore, in a certain sense, the two main treatises of special Christian ethics deal not with two separate, discrete realities, but with the one and only human reality, looking at it both in itself and as the revelatory form of man's relation to God. We can put this in another way, and give it a more accurate theological formulation: special Christian ethics examines man's relation to God, first in itself, and then as concretely realized in men's relations with other men.

II

The Three Divine Virtues

"We give thanks to God always . . . remembering before our God and Father your work of faith and labor of love and steadfastness of hope in our Lord Jesus Christ." In 1 Thessalonians 1, 2. 3 is found the earliest and oldest Christian text in which faith, hope, and love are mentioned in one breath. In Paul's letters the trio occurs several times. No text is more familiar, nor does any name these three virtues so clearly and specifically as the final lines of Paul's great hymn in praise of love: "So faith, hope, and love abide, these three; but the greatest of these is love" (1 Cor. 13, 13).

In 1916, R. Reitzenstein proposed the hypothesis that Paul formed the trio by deliberately omitting one element of a gnostic quartet.[91] He was later supported by R. Bultmann,[92] but the majority of authors take the view of A. von Harnack, and are convinced, even simply on the basis of comparison of numerous other Old and New Testament expressions, couplets, and combinations of three, that this Pauline trio—faith, hope and love—is somewhat arbitrarily chosen to formulate Christian existence. For the actual meaning of the words "faith" and "love" in many texts is identical, and the relation to God indicated by these words implies hope and

[91] *Historia Monachorum und Historia Lausiaca* (Göttingen, 1916), pp. 100–102; also cf. pp. 239, 242–255.

[92] "Ginósko," in *Theol. Wörterbuch z. N.T.* 1, p. 710, n. 78.

expectancy. This is even stated explicitly in 1 Peter 1, 21: ". . . your faith is at the same time hope in God."[93]

Although Paul's trio shows a certain arbitrariness in the choice of words, still, to a certain extent, it is an adequate definition of the life of a Christian. For in Paul's writings particularly, besides the relation to God and the expectation this implies (expressed in faith and hope), the Christian life includes also a relation to fellowmen expressed by preference in the term "love," as noun or as verb (Gr., *agape, agapao*). In this respect there is a fairly clear difference in nuance between the two New Testament authors in whose writings the term "love" chiefly occurs. In Paul, love is chiefly love of fellowmen. In John, love of God is expressed in the love of fellowmen, but his use of the word is much wider in its significance than Paul's.

Several points reveal remarkable implications when compared with the later theological tradition. The faith-hope-love trio remains inseparably together (and in this Augustine's influence certainly played its part). About the beginning of the 13th century, faith, hope, and love began to be qualified as *"virtutes theologicae"* (theological, or divine, virtues). In the language of the 13th century, their object as divine virtue is God himself, and in this way they are distinct from the cardinal virtues which relate to man and to the world.

Dating from its appearance in the *Sentences* of Peter Lombard,[94] the distinction between faith-hope-love and the four cardinal virtues has been maintained by tradition down to our own times.

Comparison of this tradition with the Pauline texts, which are its primary basis, will reveal that a distinction corresponding to that between the divine and the moral virtues must be made between faith and hope, on the one hand, and love, on the other. For, in the Pauline texts, faith and hope refer to man's relation to God, whereas love refers to his relation to his fellowmen.

Thus, when the Pauline trio is qualified in the tradition as "divine virtues," the Pauline content of the word "love" has actually been

[93] Various Dutch and German translations of the New Testament opt for this way of bringing out more explicitly the underlying identity of meaning in the words "faith" and "hope." The Revised Standard Version, the Jerusalem Bible, and English translations, in general, render this passage: ". . . so that your faith and hope are in God."

[94] 3 *Sent.* 23–32, and 33.

replaced by a particular meaning as used by John, and the content of Paul's meaning of love has been classified among the cardinal or moral virtues. In other words, Paul's trio contains not only the three divine virtues which *appear* to be based on texts from Paul, but also the four cardinal virtues, which are practically identical with all that Paul means by "love."

The matter becomes still more complex when we consider that in St. John particularly, the same word (*agape,* etc.) expresses both love for God and love for fellowmen. Thus, this love signifies what the later tradition calls "divine virtue" as well as what it calls "cardinal virtue." The complication lies specifically in the fact that some writers employ the word "love" (charity) as a divine virtue, without any distinction between love of God and love of man.

Further investigation of these complications is unavailing. We have seen enough to realize that similarity in the use of words can be highly misleading and that the fundamental rightness of the distinction between divine and cardinal virtues, as employed by theologians since the 13th century, is verified primarily by reality and not by words.

We can now return to the discussion of man's relation to God, or the divine virtues. Taking for granted the situation sketched briefly above regarding the terminology, I shall restrict my remarks chiefly to the term "faith" in discussing the reality.

The Netherlands Catechism of 1948 defines faith in this way: "The virtue of faith is a supernatural gift of God, by which we accept as true all that God has revealed" (q. 392). I think it important to state at once that I regard this definition as correct and to the point. However, it is a formulation with its own history, and it is necessary in our context to say something about its history.

Man's relation to God is represented in scripture as a deeply personal one, and comparisons are drawn with the intimacy of family relationships: with the relationship of father and son, with friendship, with the relationship of man and wife, bridegroom and bride. Scripture refers to this intimate and personal relationship in many terms, and faith is one of the many.

Behind this notion of faith is a Hebrew word which has the connotations of "to make to stand firm," "to make to be," "to establish," "to recognize"—not theoretically, but practically, in the concrete, and existentially.

To understand the full sense of this faith, however, we should realize that scripture is not merely describing the *nature* of this relationship to God, but is, above all, proclaiming the *existence* of this relationship. Scripture reveals *the fact* that man—every man, and not only the Israelite—*is* related to God because God relates himself to every man. There is, perhaps, scarcely any other text in scripture that draws this out as tellingly, as mercilessly, and yet, at the same time, as mercifully as the little book of Jonah, few traces of which we find in the New Testament, but those few go very deep. The prophet, as we know, refuses to go to Nineveh, and later in the story the reason for his refusal becomes clear. It appears he is not happy about Nineveh's conversion and its escape from the calamities which God has planned. The book of Jonah is thus a poignant indictment of the Jew's tendency to claim a monopoly of God's love for men. It is an indictment and also a proclamation, and the climax is reached in the question that ends the book: "And should not I pity Nineveh, that great city, in which there are more than a hundred and twenty thousand persons who do not know their right hand from their left . . . ?" (Jon. 4, 11).

Man's relation to God is not a mystical accomplishment of chosen souls, but a fact and a reality which confronts every man in his own personal human situation, and simply because every man is confronted with God in his fellowmen. Therefore, as Paul points out, this faith is the kind of faith that comes to reality only in love for one's fellowmen (Gal. 5, 6). Without the works of love, faith is dead and empty. James' letter leaves the matter in no doubt whatever (Jas. 2, 14ff.).

How was it possible, it may be asked, to move from this practical and existential faith to the intellectualistic "acceptance of all that God has revealed as the truth"? The simplest answer would be: through theology. This answer is a little too much of an oversimplification—if that needs to be pointed out. But a brief view of the important phases in theological reflection and in the development of the Church is better than any oversimplified generalization.

The first factor giving rise to the intellectualist interpretation of faith is found already in the New Testament and in the earliest writers in the Church. They were led to speak of a knowledge of God given to *everyone,* in reaction against the pretense to secret knowledge by the initiates of the gnostic sects. The knowledge of which the early Christian writers spoke was clearly the practical and factual knowing or recognizing of which scripture always speaks, but the words "knowledge" and the Greek *"gnosis,"* lifted out of their context by readers who were no longer familiar with their historical connotations, could easily be understood in a one-sided and intellectualistic sense.

Obviously, the meeting of the Christian proclamation with the world of Hellenic culture would make this tendency all the more acute. For indeed, the Greek word for faith, *"pistis,"* was a technical term in Greek philosophy, signifying "opinion," the lowest degree of knowledge and far removed from genuine scientific knowledge. Clement of Alexandria took great pains to make clear that Christian faith was something entirely different from this philosophical "opinion" and that it far exceeded scientific knowledge. Now, in spite of the careful intent of one of the greatest theologians in the Church's tradition, what was more likely than that faith would come to be understood as a better sort of scientific knowledge?

Conclusions of this kind are quite foreign to the biblical and early Christian authors. When we come to some of the later writers, however, there is good reason for asking about the sense in which they understood the texts of their predecessors and whether they did not initiate that intellectualizing deformation which in other places and later ages would become more and more commonly accepted. In this connection, we have to examine authors like Evagrius of Pontus (346–399), Gregory of Nyssa (c. 330–394), and Pseudo-Denis (writing probably between c. 485 and c. 518) who exercised a tremendous influence on Eastern and Western monasticism and, therefore, also on the mentality and spirituality of Western Christendom. We must also examine such writers as Maximus the Confessor (c. 580–622) and John Damascene (c. 675–749) who directly or indirectly became known in the West from the 12th century onward and, to some extent, already in late Carolingian times through the influence of the remarkable and famous John Scotus Erigena (c. 810/15–after 870). All these roads and channels are relevant if we want to understand how the gradual change to an intellectualist view of faith came about in the West.

Besides these, other factors played their part. From the very beginning of Christianity, a rudimentary confession of faith was coupled with the ceremony of baptism. Under the influence of the great trinitarian and christological controversies of the first centuries, these rudimentary formulas developed into the official ecclesiastical "symbols," signs by which the faith could be recognized. These are well known today in the form of the Nicene (Councils of Nicaea and Constantinople) Creed and others. Thus, to believe, or "to accept the faith," could be taken simply as acceptance of the official ecclesiastical creeds.

This combination of concepts of faith quite certainly had an influence on the later Latin and Western tradition. Some questions remain, however. For example, what precisely did Augustine, in spite of himself, contribute to this, not only by the use of such terms as *"intelligere," "cogitare,"* and so on—terms which became current in later theology—but, in particular, by his significance for the Latin cultural heritage of the West? For in this heritage the central concept *"auctoritas"* and the correlative concept *"credere"* in their origins certainly had no intellectualistic content (*"auctoritas"* is more or less synonymous with "authority of rule" as opposed to "authority of force"; and *"credere"* is "to cast oneself entirely upon").

In the 12th century a controversy arose between Abelard and Bernard which seems to have influenced the medieval treatise on faith decisively in an intellectualistic direction; nor has the post-medieval tradition entirely escaped from this. This time, it appears, the controversy was the result of a real misunderstanding—and this only adds to the tragedy of the whole matter.

In speaking of faith, Abelard used the term *"existimatio"* to indicate that faith is not the same as the knowledge that comes from direct experience (*"cognitio"*). Bernard, however, thought that in *"existimatio"* he recognized a term taken from the Aristotelian doctrine of scientific knowledge and, therefore, that it implied a relegation and depreciation of faith to the level of human "opinion" (cf. Boetius' *"opinio vehemens"*). We say Bernard, but it may also have been William of St. Thierry who acted toward Abelard as a sort of personal inquisitor and denouncer, often shielding himself behind Bernard. However, Bernard and Hugh of St. Victor and many theologians with them and after them insisted that faith is a "certitude" and not merely an "opinion." They went further and gave this "certitude of faith" a place in the Aristotelian scale of *"status mentis"* ("states of mind"), between "opinion" and "scientific knowledge" (*"supra opinionem et infra scientiam"*). As a result the concept of faith was, as it were, closed up for a long time to come in a one-sidedly intellectualistic framework.

Still another factor had contributed to the same result before the advent of Scholasticism. In Scholasticism this acquired a special force because in an intellectualistic framework, when the question of faith came to be developed as a separate treatise, it became more and more autonomous and less and less related to other questions. This other factor can best be described in modern terms as the divorce of faith from its concrete, everyday, horizontal context of love and common humanity. The pure "verticalism" that became increasingly more typical of the view of faith had many causes. Among these were those which had led to the intellectualistic view of faith as well. In a certain sense, this verticalism reached its most explicit form in the Reformation and in Reformed theology, but Reformation and post-Tridentine Catholic theology both have their roots in late Scholasticism.

Nevertheless, from the very beginning, the Reformed view of faith was characterized by an extremely personalist accent. This did not appear in Catholic writings until the end of the 19th and into the 20th century, when it began, explicitly and systematically, to be included in treatises on faith, even though the terms in which it appeared were often fairly vague and therefore subject to just criticism.

Catholic authors in recent years have produced a number of publications on faith which can be termed typically Protestant because of their verticalism, although it is possible that the authors concerned are not aware of their actual historical position. Both personalism and the intensified verticalism of some recent books are a reaction against the one-sided intellectualism of the view which holds that faith is the simple acceptance of truths revealed by God.

A third way is open, however, in which it is possible to do justice to the various tendencies in the tradition without falling into an unbalanced intellectualism or verticalism. Examples of this third way can be found in a number of sources, among them the *Decree on Justification* of the Council of Trent, the Constitution *Dei Filius* of Vatican Council I, and the *Constitution on Divine Revelation* of Vatican Council II. These particular texts stand out not primarily because of their theological and theoretical excellence, but, above all, because of the great many-sidedness which they, in fact, reached after much discussion and many amendments. Many-sidedness is not lacking in the theological tradition, but there one often finds it outside the treatise on faith, usually in the treatise on love and on the cardinal virtues.

In the first article of the *Secunda Secundae,* Thomas, without any restrictions or one-sidedness, gives an extremely penetrating definition of faith as a relation to God. We have already noted, in connection with the text of Jonah and elsewhere, that scripture reveals primarily the concrete existence of every man's relation to God. The same thing is found in Thomas in the fact that the treatise on faith opens with a question about the object of faith. As we go on, it will become clear with what deep insight Thomas here employs the principle *"actus* [*et habitus*] *specificatur ex obiecto."*

Thomas chooses the formula *"obiectum fidei est veritas prima"* which had been used by William of Auxerre, but he gives it a very particular interpretation. Just as in every case of Thomas' use of *"primum,"* he wants *"veritas prima"* understood according to the analogy of the *"ratio formalis obiecti."*[95] Let me explain. The *"primum visum"* is that by which a thing can be the object of the faculty of sight (*"primum visum"* being "color"); the *"primum auditum"* is that by which a thing can be the object of hearing (*"primum auditum"* being "sound"); and so, too, with *"primum cognitum"* and *"primum volitum,"* and so on. That which we see, primarily, is color, and not a particular thing; the hearing hears sound primarily, and not a particular determinate sound. In the same way it is the Divine Truth that we primarily believe, and not a particular truth, such as the humanity of Christ or the sacraments of the Church.

However, one must be very careful to understand this. In a certain sense, faith is not a reality apart and in itself, but it is entirely brought about by God as *veritas prima.* This means that faith is brought about by God *as manifested in* that which actually confronts man. "By faith we believe what God has revealed": this is the traditional formula, and Thomas uses it in the first article of his treatise. It does not mean "I believe the things God reveals," nor does it mean "I believe them because God reveals them." Its primary meaning is this: "God reveals himself *in* his world, and

[95] Cf. *Tijdschrift voor Theologie* 4 (1964), p. 167, n. 54.

because of this I see that my involvement in the world is involvement with God himself." This is faith.

There can be no doubt about Thomas' view of faith, and this can be shown both in his treatise on God, where he explains the all-embracing significance of *"ipsa summa et prima veritas,"*[96] and in his treatise on faith, where he explicitly includes *"quaecumque creaturae,"* the whole of created reality.[97] He gives the name of faith to the totality of concrete human existence, not indeed precisely as an earthly reality, but precisely and exclusively insofar as every man is confronted in everything, everywhere, and at all times by the God who manifests himself in man's world and who is concerned with the lot of men who, by themselves, do not know the difference between their right hand and their left. In other words, faith itself is an aspect of the mystery, of the manifestation of God in humanity, and is first revealed by the proclamation of the mystery.

Thus, in the whole tradition of faith, hope and love—that is, of the divine virtues—we actually and fundamentally remain within the line of the faith that becomes reality in the love of our fellowmen (according to Gal. 5, 6) and that, in this extremely concrete sense, really does accept all that God reveals.

Besides this fundamental and decisive sense of faith and of believing, there is good reason for speaking of faith and of believing in a more limited sense, with reference to those things mentioned in the salvational proclamation and the Creeds. But if this faith is not an accompaniment to and clarification of our practical and matter-of-fact recognition of God in our fellowmen—our non-mystical neighbors who hardly fit the mold of our wishes and desires—then it is an empty cry of "Lord, Lord," a lie, a lip service, a dead faith (Jas. 2).

It is probably superfluous to add here that the famous love of neighbor "for God's sake" can have no meaning other than that intended above. In other words, here, too, the primary fact is love for man, for this love for man is revealed to be love for God; consequently, love cannot reach God unless it reaches man, as man, for himself. The "neighbor" is not an unavoidable steppingstone on the way to God; he is the *"imago Dei,"* God's human

[96] *Summa Theologiae* I, 16, 5.
[97] *Ibid.*, II–II, 1, 1 ad 1.

form. This reality provides the primary foundation of actual faithfulness or unfaithfulness.

III

The Four Cardinal Virtues

While the first part of special Christian ethics dealt with the relationship of man to God in itself, the second part views this relation to God precisely insofar as it is realized in common humanity, and thus deals with the mutual relations of men. In Pauline terms, one would have to call this the treatise on *agape,* love (of neighbor)—provided that one did not understand the word to mean a sentimental friendliness, but the entire complex of all relations of men with one another precisely as the incarnation and form of God's own love for men. All but a very few instances in the tradition of moral theology, however, place this treatise under the title of the four cardinal virtues: insight ("prudence"), justice, fortitude, and moderation (*prudentia, iustitia, fortitudo, temperantia*).

In this context, the history of this treatise is of importance only insofar as it is needed for an understanding of the sense of the discussion and, later, for clarifying the problem of conscience.

The history of this quartet of virtues, as far as we know, begins with Plato. Plato's distinction of four virtues is not arbitrary, but is based on an anthropology that held there were three "parts of the soul"—understanding, desire, and spirit. To these correspond the three virtues of wisdom, moderation, and fortitude. Justice is the harmony of these three parts of the soul and is thus, as it were, the product of the other three virtues. In this sense, justice is central to all virtue and, in a certain way, the sum of all virtue.

To the extent that one may distinguish Aristotelian anthropology as extravert from Platonic anthropology as autarchic, the replacement of wisdom by practical insight is a symptom of the difference between the two great philosophies of ancient Greece.

However this may be, the fact remains that the four virtues entered the various later stages of Greek philosophy, particularly among the Stoics, with thoughtfulness (Gr., *phronesis;* L., *prudentia*) and not wisdom (Gr., *sophia;* L., *sapientia*) alongside the other three, moderation, fortitude, and justice. Although it is agreed that this already represents a certain Aristotelian influence, the later stages of history show that the difference is largely one of terminology, and that not too much importance is to be attached to it.

The manner in which the Christian tradition employed this quartet of virtues seems to me to be very significant theologically—more significant, in fact, than all the various concrete ethical conclusions that were attached to the quartet with greater or less explicit connection in the course of the centuries.

The four virtues are mentioned only once in scripture—in Wisdom 8, 7. As far as I know, no doubt exists among the experts that the author of Wisdom (writing in the metropolis of Alexandria around the middle of the first century before Christ) adopted the four virtues from what is usually known as popular Greek philosophy. His personal and only contribution in this respect is that he presents these virtues as the fruit of Divine Wisdom and of the justice that God himself engenders among men.

We find fundamentally the same thing in Clement of Alexandria. He, too, has one objective: to take over the current description of virtue as it is, in order to establish that these virtues are gifts of God.

The Western philosophical tradition was first influenced by Cicero (106–43 B.C.) who gave Greek philosophy a Latin tongue and translation. The four virtues are introduced to the West in his book *De Officiis*. In the 4th century, Ambrose wrote a book under the same title, on the same subject, following the same lines as Cicero for the most part. No indication has been found to show that Ambrose was inspired or led by the text of Wisdom 8, 7. But it must be stated that Ambrose was the first to name these four virtues the "cardinal virtues" and that it was only in the 12th century, through the school of Laon and especially through the work of Peter Lombard, that this became the accepted term in the theological tradition.

Augustine, at a fairly early date, certainly learned of the four virtues in Cicero's *Hortensius,* for which he had such great admiration. He often mentions and discusses the four classical virtues, but in his *Retractationes* he states that it was only much later that he discovered that they were named in Wisdom. This does not imply, of course, that Augustine did not know the Book of Wisdom before. There are other reasons for his not knowing, and his remark in the *Retractationes* is therefore quite revealing with regard to Latin culture. In the beginning, Augustine had known only Latin translations of the bible. Among these, one had a fairly significant variation of Wisdom 8, 7, and three others translated the four virtues in terms other than those which Cicero had used.[98] Because of this departure from what Augustine spontaneously regarded as the standard Latin formulas (with which the bible translators were obviously not familiar), he did not recognize them until he obtained a Greek text of the scriptures and noticed that the four virtues, of which he had so often spoken, were also mentioned in the bible.

Augustine's case thus sharpens our awareness of the fundamental theological fact which we have already noted in Ambrose, in Clement of Alexandia, and in the Book of Wisdom itself: the Christian proclamation does not bring a new ethic of its own, but reveals the salvational and salvationless character of the human ethos. The Christian message is not a message of new and special virtues taken from scripture, but an illumination of the divine character of the concrete, human, cultural ideal.

In light of the previous chapters in this book, this idea is, in a certain sense, not new. What is new is the fact that the actual history of concrete ethics confronts us once more with this fundamental Christian truth. This serves as yet another protection against incorrect opinions and conclusions about the history of ethics. In many recent publications of a more or less theological nature, and in some that are not so recent, criticism is brought to bear on ethical attitudes of the past. Usually this makes sense only in terms of the presupposition that these ethical attitudes are just as valid now as they were then. If one does not acknowledge this validity, the only possible alternative is to hold that the attitudes of the past were wrong. In consequence, all sorts of people, like the Greek Fathers, Augustine, Gregory, Thomas, and the Scholastics must be accused of being in error, and apparently the only competence required of one in order to be able to do this is the right to express opinion freely.

Theologically, this is uncalled for; humanly speaking, it is hardly edifying. But what is stranger still and, in a certain sense, more discouraging is the fact that disputes waged in this way are so often

[98] "*. . . sobrietatem enim et sapientiam docet, et iustitiam et virtutem.*"

waged against the very presuppositions and projections of the disputants themselves. The ethos of other cultures and other times cannot of itself have any perpetually normative significance, even where it forms part of a theological tradition. The theological tradition itself *is* normative, but it is normative precisely, primarily, and formally as a theological tradition and not as an ethical one. In other words, in many cases, including the case of the four cardinal virtues, tradition bears constant witness to the salvational and nonsalvational significance of the concrete human ethos; in so doing, it canonizes no particular ethos, and certainly no particular systematization of ethics, or makes it into a perpetually valid rule.

The facts we have found in the first few pages of the history of the four cardinal virtues within the Christian tradition are confirmed in many ways by the rest of this history. The only reason for going into this matter more deeply here and now is that it will help to localize and clarify the problem of conscience more easily.

The Latin Fathers after Augustine, the Carolingian theologians, the early medievals, and the Scholastics all knew and employed the classical quartet of virtues without giving it any extensive or remarkable attention. It was only in the 13th century that concrete ethics began to be systematized around the four cardinal virtues. Thomas took this line and worked it out logically and at great length, but clearly aware of the relative value of the scheme he was following. Thus in fact the ethics of Thomas are entirely centered on the treatise on justice—that is, on the demands of others.

This remained the central point throughout the post-medieval development—whether the scheme of the four virtues was used or not, and, if used, whether for deliberate reasons or merely *pro forma*. Thomistic authors clearly felt themselves obliged to maintain the scheme. But the deeper the authors go into practical morals, the greater the chance that they opt for other schemes. In this respect, the famous manual by the Jesuit Busembaum, predecessor of Alphonsus' moral theology, was a pioneering work, and within 100 years after its appearance in 1650 (possibly 1645) it had gone through 200 editions. It will be obvious that any scientific criticism would have been wasted energy. If there is anything to bemoan in this development, it is probably not so much the inflatory influence of practical morals on the classical treatment of the four cardinal virtues as the remarkable upsurge of interest in conscience.

IV
Conscience

The French Dominican school, and particularly Garrigou-Lagrange, Noble, and Deman, in opposition to the practical moralists, defended the thesis that Thomas deliberately left the question of conscience out of his chief theological work in order to replace it by prudence or practical insight. We no longer encounter the noticeably fervent supporters of this thesis, but the notion that Thomistic morals are morals of prudence and not morals of conscience has gained a wide circulation.

This notion certainly is based on an obscurity, and its clarification is very important in the light of the great interest attached to conscience and "acting according to one's conscience."

The concept of conscience (Gr., *suneidesis;* L., *conscientia*) arose out of particular ethical views in the Graeco-Roman world around about the time of Christ.[99] In this ethical context, which was not entirely free of a measure of egocentricism, "conscience" had more or less the sense of self-consciousness, personal judgment, personal conviction, self-sufficiency, independence, and so on. "Conscience" was a term expressing a human ideal of personal completeness and perfection of the kind that could not be attacked or disturbed by the inevitabilities of life, the opinions, views, and customs of others. To follow one's own conscience and to act according to one's own conscience in this context means not to allow oneself to be influenced by others and to go independently along one's own way, regardless of what others may say or think.

The characteristics of this self-conscious attitude to life are nowhere so well described as by Paul in 1 Corinthians 8–10. In the community at Corinth, Paul found himself confronted with certain problems which, at least to some extent, had been caused by Christians who obviously adhered to this ethic of conscience. They had come into conflict with other members of the community on the matter of eating meat that had been sacrificed to idols, and

[99] Cf. especially J. Dupont, O.S.B., "Syneidesis. Aux origines de la notion chrétienne de conscience morale," in *Stud. Hellen.* 5 (1948), pp. 119–153; bibliography according to Maurer in *Theol. Wörterb. z. N. T.* VII (Stuttgart, 1964), pp. 897–898.

that meant practically all of the meat that one could buy on the open market.

The others judged that, as a Christian, one could not eat such meat because, by doing so, one would, in fact, be taking part in the worship of false gods and giving it public approval. They based themselves, however, on the principle that there were no false gods and, therefore, that offerings to false gods had no meaning; consequently, they held that there was no reason whatever to abstain from sacrificial meat which, after all, is and remains a gift from the one and only God and creator who is their Lord.

In this part of his letter, Paul uses several phrases that express this ethos of conscience more precisely. The Christians concerned have obviously claimed that they, unlike the others, have "knowledge" and "insight" (cf. 1 Cor. 8, 1. 2; etc.); they "know" that false gods do not exist; they are "free" and independent (cf. 1 Cor. 9, 19; 10, 23. 29); they are "sensible" (Paul uses the word "*phronimos*": 1 Cor. 10, 15). Paul thus plays on the meanings of the popular definition of "*phronesis*" (insight), the knowledge of good and evil, and clearly his meaning is that they are convinced that they know what is good and what is bad, what may and what may not be done; they judge for themselves (1 Cor. 10, 15); they are the strong ones, unlike the others who are "weak" (1 Cor. 8, 9. 10).

Paul grants that they are right to a certain extent. There are indeed no false gods, and on these grounds there can be no objection to eating the meat offered them.

He even warns their opponents. Keeping the law neither justifies nor leads to salvation. For, indeed, all were under the cloud, and all passed through the sea, and all ate the same manna and drank from the same rock, Christ, and yet many of them perished (1 Cor. 10, 1–13). In the careful fulfillment of the law, all they had done was to create a new idol for themselves. Neither will the Corinthians escape from idolatry if they avoid sacrificial meat and, in consequence, make fulfillment of the law a new idol.

Nevertheless, in all this Paul is at the same time attacking the ethos of conscience, at least in its concrete results. For one is able to make a god of one's own conscience, too, by following that conscience and acting on one's own principles to the detriment of one's fellowmen. This is the basic fault of which Paul accuses them. In their abstract and egocentric concern for theory and principle, they do not see what is really at stake and that ultimately it is not a matter of theory or of principle, but quite simply of the person standing next to them.

In contrast to their kind of liberty, Paul boasts that he is the slave of all and that his freedom consists in being of service to those who accept the law and those who do not, to Jew and to pagan (1 Cor. 9, 19ff.). It is all the same to him, he writes in another place, if people celebrate certain feasts or do not celebrate them, if they see anything in the new moon or not, if they want to be vegetarians or to eat meat, as long as they do what they do to the honor of God (cf. Col. 2, 16; 1 Cor. 10, 31; etc.), and the unmistakable criterion of God's honor is their attitude toward their fellowman.

In a certain sense, then, Paul is also indifferent to the ethos of conscience, although salvation and the concrete interests of concrete persons in Corinth make him urge them to act "for conscience' sake—I mean, his [your neighbor's] conscience, not yours" (1 Cor. 10, 28–29).

Paul does not oppose the ethos of conscience as such in principle, but only the egocentricism that was, in fact, connected with it in Corinth. It will have to be admitted, however, that this egocentricism is inherent in any ethos of conscience, and in this sense Paul's criticism is more extensive than the particular conflict with which he had to deal in the community at Corinth. It seems to me that history has proved this and that, for precisely this reason, Thomas' position is so significant and can clarify so much.

The problem of conscience, so keenly formulated in the 12th century by Abelard (the Corinthian party of which we have been speaking would certainly have applauded, at least, before Paul tackled them), constituted a very important part of medieval theology. The central questions in this problem were questions of the erroneous conscience: If one acts according to a conscience that is in error, does one do well or does one do ill? Can the error be corrected, and must it be corrected?

Taking note of Thomas' usual attitude to all the theological questions of his time, it can surprise no one that he brings the problem of conscience into the discussion several times in the *Summa,* wherever his context requires it. Apart from a number of incidental remarks or answers, however, Thomas gives particular attention to the problem only twice, and briefly at that.

The first of these two occasions is when he considers conscience among the acts of the understanding[100] along with the mysterious *"synderesis,"* which actually seems to be nothing more than a copyist's error for *"suneidesis"* in one of Jerome's texts.[101] The second occasion is when Thomas gives the question about erroneous conscience a place in the discussion of questions about good and evil.[102]

The thesis that Thomas replaced conscience with prudence or practical insight[103] is based on a number of considerations. We need not detail them here. A certain weight is given to the fact that Thomas classified conscience as one of the acts of the understanding and that among the virtues he deals with *"prudentia"* or practical insight, but not with conscience.

In a criticism of this thesis, one of the first questions to be asked would be whether it is constructed upon an examination of Thomas' own work or upon the work of his commentators. It seems more likely that the thesis is drawn by the French Thomists from the great treatises on conscience in the later manuals of morals than from the actual situation in the 13th century, for there the problem of conscience played a less important role than it played in the so-called fundamental morals of the manuals of the 19th and 20th centuries.

Another point is much more important, however. It seems that Thomas did quite deliberately undertake something that was correctly identified by the French Thomists, all of whose implications, however, they did not see. It is a fact that conscience receives Thomas' attention only in a most modest fashion, but this was not because he chose to have *prudentia* in its place, but rather because he could deal with all the questions arising in the problem of conscience in the treatise on the four cardinal virtues. In general, too little attention has been given to Thomas' own sense of history, with the result that it is too little appreciated. With this sense of

[100] *Summa Theologiae* I, 79, 13.

[101] Cf. M. Waldmann, "Synteresis oder Syneidesis? Ein Beitrag zur Lehre vom Gewissen," in *Theol. Quartals.* 119 (1938), pp. 332–371; J. de Blic, S.J., "Syndérèse ou conscience?" in *Rev. d'Asc. et de Myst.* 25 (1949), pp. 146–157.

[102] *Summa Theologiae* I–II, 19, 5 and 6.

[103] This seems to me to be the obvious implication of the thesis of the French Thomistic school: that subsequent theology replaced "prudence" with "conscience."

history, Thomas recognized that in the problem of conscience he had to deal with a thematic ethic that was different from the one represented in the scheme of the four cardinal virtues. The radically egocentric character of conscience made it considerably less suitable as an ethical framework than the classical scheme of the four virtues. For this reason Thomas chose the latter and not conscience for the framework of his ethics. Yet, it is clear in many texts that he was very well aware of the content of the ethics of conscience and that he had thoroughly understood it.

It does not fall within the scope of this present work to detail the grounds for a particular interpretation of Thomas. But a closer and more practical discussion of the problem of conscience in itself does seem to be called for, especially in view of the actuality of the notion that one must "act according to one's own conscience."

The history of the 20th century, just as much as that of the 1st or the 13th, tells of the egocentricism that, time and again, is seen to be inherent in the ethos of conscience and in every reference to "acting according to one's conscience." From the earliest beginnings, conscience has been the purely personal affair which fits very well indeed into the Stoic striving after imperturbability, yet is extraordinarily difficult to harmonize with the essentially human task of communication and community, with living, loving, and suffering along with fellowmen who are all too readily perturbed and disturbed.

Probably the best thing, therefore, would be to discard altogether this radically and essentially vitiated concept of conscience. The fact remains, however, that in practically all languages, at least in the Western world, the word and the concept are in common use, and this is a fact that would indeed be difficult to ignore. The only realistic solution, therefore, lies in a very careful examination of the reality and in formulating the reality, if need be, even in terms of conscience.

Remarkably, the tradition of moral theology, driven by every kind of question, has already done this. For, indeed, it speaks not only of conscience, but also of a just conscience and an erroneous conscience, of the principles and laws of conscience, of the voice of conscience, and the like. Formulas of this sort reveal the ambiguity that is inherent in conscience as soon as it departs from its own proper and egocentric context, for out of its egocentric context conscience must express two elements: the personal judgment of the situation and the concrete demands of the human situation. These two elements together may well form one single reality, but the single word "conscience" is not able to give adequate expression to both aspects of this one reality. The often-heard objections to conscience or to actions which are "according to conscience" are not directed against the personal judgment or conviction of the one who so acts; rather, they are directed against the fact that conscience here so often means this personal judgment only, *to the exclusion of* the objective demands of the human situation.

If one considers the fact that moderation and fortitude are, in a certain sense, parts of justice, one can readily understand the preference for the scheme of the four virtues over that of conscience, for it is precisely the ambiguity of conscience that the other scheme eliminates. The personal judgment is assumed in practical insight, and in justice the objective demands of the human situation are brought to the fore with the greatest emphasis. Thus the French Thomists themselves were, in fact, sidetracked into the subjectivism of conscience with their rather one-sided thesis about *"prudentia."*

A second reason why the scheme of the four virtues is to be preferred to that of conscience is found in the distinction between conscience as knowledge, and practical insight, justice, and so on, as virtue. Human action does not proceed from knowledge, but from habit or virtue. The theorists of conscience themselves acknowledge this insofar as they do, in fact, present conscience as a virtue.

A third reason argues against conscience and for virtue. Conscience, with its principles and laws, belongs in the sphere of the "social language"—that is, it belongs in the sphere of code, fixed rules, and norms. There is a certain validity in this, but it does not reach to the fundamental reality of personal communication and of response to personal desires and demands.

The scheme of the four virtues has practically disappeared from our everyday surroundings. One can get along, if need be, with conscience, but only if one gives his full attention to things that actually are foreign to the field of conscience. The interpersonal relationship could provide the new framework for a concrete ethic, for, indeed, it covers the same field as the love of which Paul speaks and the justice of the four-virtue scheme.

The titles and formulations actually chosen for a special Christian ethic constitute a secondary question, the more easily answered as the real content of Christian ethics is more clearly perceived. This content can be described as the divine and cardinal virtues, but also in other ways: Church and world; the relation to God and the relation to fellowmen; faith and brotherly love; faith in God and in fellowmen; love of God and the interpersonal relation; justification and justice; grace and humanity; divine community and human society; salvation and humanity; and so on. Conscience, however, is better left under the dust of the ages where it really belongs.

2. Religion and Prayer

I

Familiarity with the God Who Became Man

A whole sea of literature exists on religion as well as on prayer. If this chapter refers to it at all, it will be only in the most general terms. What follows will be based more exclusively than all that has gone before on an analysis of Thomas' texts, for, in comparison with a great deal that has been written on this subject, they recommend themselves by their fundamental and radical clarity.

Catholic theologians have repeatedly offered objections to the placing of the treatise on religion and prayer within the context of the cardinal virtue of justice. In their opinion, religion and prayer belonged among the theological virtues, and they often directed their protests against Thomas explicity. For he and those who follow him have considered religion and prayer in connection with justice.

Oftentimes these modern authors do not realize that they are formulating objections that had been previously discussed in the 12th and 13th centuries. Certainly since the time of the Stoics and of Cicero, *"eusebeia,"* or *"religio,"* has been connected with *"dikaiosune,"* or *"iustitia."* Such medieval authors as Alanus of Lille, Philip the Chancellor, William of Auxerre, Bonaventure, Albert, and Thomas found this a problem: Does religion belong here or among the divine virtues? At first, their answers were diametrically opposed. Alanus, for instance, regarded faith, hope, and love as parts of religion; Philip appears to agree with him entirely. On the other hand, William of Auxerre thought that religion could not be reckoned a divine virtue.

Bonaventure, in giving a more precise definition of religion, assumed a kind of middle position. Religion as external worship must be classified under justice, but, as an inward worshipful attitude it belongs to the divine virtues. Inasmuch as the proper meaning of religion refers to outward worship, religion as such belongs simply to the cardinal virtue of justice. Albert accepted this opinion, apparently without question.

Thomas agrees with his great predecessors. He, too, holds the distinction between outward worship and the inward which consists in faith, hope, and love.[104] Therefore, the outward worship must be called the expression of faith (*"protestatio fidei"*).[105] However, when Thomas comes to deal with religion explicitly (and he does this in the treatise on justice), he does so more fundamentally than those who went before him. Thomas speaks, first of all, of the virtue of religion,[106] then of its so-called inward acts (*actus interiores*) which are devotion (*devotio*)[107] and prayer (*oratio*),[108] and, finally, of its outward acts.[109] What to our modern ear is the primary connotation of religion (the outward manifestation) is, in Thomas' order, the last. A similar fundamental approach had already, in the treatise on faith, led to remarkable results. What is the significance of his order with regard to religion and prayer?

Presupposed in the entire treatise is a general definition of the concept of religion as the ordering of God to man.[110] This is no arbitrary determination, but the firm recognition of a fact in Western tradition. When we speak of religion here, we mean the order and directedness of man toward God.[111]

The analysis of religion as a virtue leads now to a thesis similar to that reached in the analysis of faith. The ordering toward God is present in man primarily as a given fact, corresponding to the actual relationship between man and God. God is creator, cause,

[104] *Summa Theologiae* I–II, 103, 3.
[105] *Ibid.,* I–II, 100, 4 ad 1; 103, 2.
[106] *Ibid.,* II–II, 81.
[107] *Ibid.,* II–II, 82.
[108] *Ibid.,* II–II, 83.
[109] *Ibid.,* II–II, 84–91.
[110] "*. . . religio proprie importat ordinem ad Deum*": *ibid.,* II–II, 81, 1.
[111] Note especially the quotations from Cicero and Augustine, *ibid.,* II–II, 81, 1, sed contra et ad 2.

and principle of all things.[112] The right relationship of man to creation and thus to his fellowmen necessarily involves the ordering of man toward God. Therefore, religion, as a reality and virtue, is not, in the first place, a reality to be cultivated separately and alone; it is, in fact, something that is implied in all virtue or in all authentic humanity as such. The religious man is not a peculiar phenomenon, as is sometimes thought and as some psychologists would have it, arguing from empirical observations and, apparently, just as much from preconceived ideas. Fundamentally and primarily, the religious man is not the one who displays outward signs of his religion, but the one who in all respects is humanly good as a man, even if he sometimes regards himself as an agnostic or an atheist. For the directedness toward fellowmen involves precisely that directedness toward God which is expected of men.

Thomas therefore holds that religion not only indicates this directedness toward God, but also that it indicates this directedness in the strict sense (*"proprie"*) and formally. The first four arguments (or objections) in the first article (81, 1) involve the fellowman in religion, beginning with the well-known text of James 1, 27: "Religion that is pure and undefiled before God and the Father is this: to visit orphans and widows in their affliction. . . ."—a theme that the prophets of the Old Testament also held up and emphasized to the nation so ready to bring its sacrificial offerings. Thomas' intention was certainly not to exclude fellowmen from religion, except in the formal and precise sense: religion indicates man's directedness toward his fellowmen precisely and only insofar as this is, in fact, directedness toward God. Religion and common humanity are materially synonymous; formally, they are distinct.

This distinction must be made in regard to actions just as it is made in regard to virtue. Every good human act is, at the same time, a religious act and can therefore be classified not only in terms of interpersonal relations or of justice, but also in terms of religion: as a sacrifice, for example. If one proceeds from the principle that religion, which is an ordering toward God, is primary and the ordering toward men is secondary, one can understand how Thomas came to formulate the above-mentioned distinction in the technical terms of the Middle Ages as *"actus eliciti"* and

[112] Cf. *ibid.*, II–II, 81, lc, ad 3, ad 4; 81, 3 and 8.

"actus imperati" of religion as a virtue.[113] Here, too, it must be remembered that *"actus elicitus"* and *"actus imperatus"* are materially one and the same act, which means that the whole of human life is either religion or its rejection.

How, finally, is religion distinguished from faith? This, of course, presupposes that they are distinct, and, as with William of Auxerre and Bonaventure, Thomas' conviction that they are distinct appears already in the mere fact of the place which he assigns to the virtue of religion in the treatise on justice.

The manner in which Thomas describes faith, religion and justice, or the interpersonal relationship, leads, as far as I can see, to one conclusion only: that which is materially one and the same human reality can be expressed as faith and as religion and as the interpersonal relation. The term used depends on that aspect of the total reality which one wishes to express: faith indicates the direct communication of man with God; religion indicates man's subjection and directedness to God; common humanity indicates man's relatedness to his fellows.

Although this already points to a certain distinction between faith and religion, the relation of one to the other remains obscure unless one keeps in mind the formula mentioned above, in which outward worship is characterized as the expression of faith, *"protestatio fidei."*[114]

One must remember that the act of belief, as such, is not an observable reality, however easily and readily such words as belief, believe, act of faith, content of faith, life of faith, confession of faith, and so on may be used. We can be believers only by the free gift of God's presence in our world. The fact of our belief, therefore, is as hidden and veiled as God's own presence and in itself remains just as inaccessible. All accessible fact, all revelation, and all observation take place, for us, on the level of the sign, *"in hysterio," "in sacramento"*—that is, it all takes place within the fields of corporeity, of the cardinal virtues, of justice, and so on. This is the reason why religion belongs in this field, and not among the divine virtues.[115]

It clearly follows from what we have said that religion can be understood in more than one sense.

[113] *Ibid.*, II–II, 81, 1 ad 1; 4 ad 2, ad 1.
[114] See footnote 105.
[115] Cf. *ibid.*, II–II, 81, 5.

1. Religion, in what may be called the more popular sense, is the particular type of activity by which man expresses his dependence on and subjection to God or the deity. Religion in this sense can also signify the *habitus* or directedness from which this activity proceeds. This recalls the manner in which the Stoics, Cicero, and many others seem to have understood religion; it is also the view of religion that comes under criticism in the Old and New Testaments, insofar as religion is conceived in contrast to and competition with man's duties toward parents, toward the weak and suffering, and toward fellowmen in general.

2. When closer reflection has revealed the error of any ideas of competition between God and man, a twofold sense of religion can be distinguished: (a) a more general sense, insofar as the whole of human life, precisely as human, is the most primary and obviously required form of religion; (b) a stricter sense, inasmuch as religion expresses the confession of subjection to and dependence on God or the deity in a special kind of activity. Once again, the "virtue" of religion is the origin of all this.

3. Within the whole framework of the proclamation of salvation in all its aspects, religion, in the more general as well as in the stricter sense, is shown to be a confession of faith, a *"protestatio fidei."* When the servant is taken into the confidence of the master (cf. Jn. 15, 15), religion, or the worship of God, becomes familiar intimacy and friendship. The awesome experience of subjection and dependence in religion becomes the expression of the close relationship that God maintains with men, and of this we have the most meaningful manifestations in the Last Supper and in the eucharist.

Thus it appears that all religion finds fulfillment in Christianity, which is the expression of the mystery of Christ, in the more general sense, because the relationship of God and men is expressed in the relations of men to one another, and in the stricter sense, because the relationship between God and men is given explicit form in sacrament.[116]

And now, finally, the relation of faith to religion can be accurately defined. We proceed from the primacy of God's becoming man and from the relationship between God and man that this entails—that is, faith and the divine virtues. In terms of this primacy,

[116] Cf. *ibid.*, II–II, 81, 7 ad 2; 89, prologue.

the entire world of human corporeity (humaneness, interpersonal relations, justice, the cardinal virtues) is sign, form, and sacrament, and the religion hereby entailed is *protestatio fidei,* or affirmation of faith. All this is due to and recognized in the light of the sacramentalization which is an explicit fulfilling and replacing of whatever is called elsewhere simply "religion." In other words, justice, or the interpersonal relation, may be qualified as "the virtue of religion" precisely as the embodiment of faith, and this virtue at the same time finds expression in the explicitly sacramental proclamation of faith and not only in everyday fellowship.

The modern term "religionless Christianity" is not a silly one because it accentuates correctly the distinction between "religion" and the concrete profession of faith. Yet, it is not a happily chosen term because it connotes an antagonism and so is unable to give proper expression to Christianity's essential fulfillment of all religion in the more general as well as in the stricter sense.

Modern authors who try to establish the close connection between religion and faith or the divine virtues are actually supported to a large extent by the medieval theologians, even though these modern authors appear to attack the standpoint of the medievals. The contradictions that some of the moderns point out arise from their own lack of precision and neglect of careful historical research.

II

Prayer

Following his discussion of the virtue of religion, Thomas goes on to discuss the acts that proceed from it, dealing first with the inward acts (*"actus interiores"*) and then with the outward acts (*"actus exteriores"*).[117]

To properly understand Thomas' view of prayer, we have to remember that he regards prayer as, in the first instance, not an outward but an inward act of religion. He distinguished two inward acts in this regard: devotion (*"devotio"*) as an act of will, and prayer (*"oratio"*) as an act of intellect. In the very nature of the case, as with all these distinctions, this is not a question of two

[117] *Ibid.*, II–II, 82, 1, prologue.

acts that are materially discrete, but of two real aspects, formally distinct, of the one human reality of the act.

The "inward act of religion" is an indication, first of all, of the actual religious character of every good human act, while the "outward act of religion" indicates concrete action that is religious in the more general sense (implicit religion) or in the stricter sense (explicit religion).

The distinction between acts of the will and acts of the intellect is likewise concerned with two *aspects* of the inward act and, therefore, with the actually religious nature of every act. Especially in connection with this distinction one must take account of the fact that Thomas employs it in order to make room for a number of traditional notions. Consequently, its significance must not be exaggerated.

Although one would think of prayer as an outward action,[118] Thomas treats it, in the first instance, not as an outward act, but as an inward act. This very fact shows by implication that he approached the subject in terms of his fundamental interpretation of religion as an aspect of human life as it is in the concrete. Note, therefore, that we are not yet considering that phenomenon which is usually and quite spontaneously called "praying"; we are now considering that particular aspect of all human activity to which Thomas wants to give the name "prayer": an aspect which precedes the outward act of praying and is the basis of it.

Perhaps it may be remarked that the whole context of the treatise on religion and the whole of Question 83 with its 17 articles on prayer clearly and sufficiently guarantee that all this is not rational theorizing, but careful consideration of the actual *"consuetudo Ecclesiae."*[119] Along with this it is necessary to remember that the precise task of the theologian is to understand and interpret correctly the actual data provided by tradition, whether academic or concrete.

The actual analysis of prayer comes in the first article of Question 83. As called for by actual practices regarding prayer in the Church of the 13th century and by the theological problems arising from them, the remaining articles add more precise clarifications of a number of points.

[118] Cf. *ibid.* II–II, 91, prologue and 89, prologue.
[119] Cf. *ibid.*, II–II, 10, 12.

Article 1 proceeds from the notion of prayer as petition (*petere*). Thomas, as well as Augustine and John Damascene, gives a definition of prayer in terms of *"petere"* in connection with various texts of scripture in which the same or a similar expression occurs. Within the framework of 13th-century terms and concepts, prayer, as an act of the intellect, belongs to the practical intellect and not to the speculative, for prayer is not a matter of mere knowing, but a matter of actually bringing something about or causing something.[120] Now this causality can be distinguished (once again, as I see it, with a formal and not a material distinction): there is both "perfect" causality (*"necessitatem inducendo"*) and "imperfect" causality (*"solum disponendo"*). The former is called *"imperare"* (to command); the latter is called *"petere"* (to request).

In the use of the word *"imperare"* (to command), the intention is to signify the actual activity of doing or bringing about—according to its precisely rational aspect—as distinct from the will and especially as distinct from the executive faculties. This can be shown most clearly from the definition of the *"imperium"* or command as a rational (and voluntary) act in Thomas' treatise on human action.[121]

In the use of the word *"petere"* (to request), the intention is to signify the element of need or potentiality which is necessarily present in every act and which is an essential component of every act. This need or potentiality itself, then, may be called "causative" with respect to the effect, although its causality is not of the perfect and efficient type, but imperfect and dispositive. Stated more concisely, all actuation presupposes not only an efficient cause, but also a potentiality. As the efficient causality can be called *"imperare,"* or commanding, by analogy the potentiality in a rational sense may be called *"petere,"* or requesting.

Thomas' analysis of prayer begins from this most fundamental aspect of human reality. In a certain sense, religion is already

[120] In the *Summa Theologiae* I, 79, 11, it is explained that *"intellectus speculativus"* and *"intellectus practicus"* are not two distinct faculties, but that the second is distinguished from the first by *"applicatio ad actum."*

[121] *Ibid.*, I–II, 17. In the treatise on human action, one can also see how Thomas in this way makes place for *"imperium"* as an historical datum that had to be accounted for; cf. *Tijdschrift voor Theologie* 4 (1964), p. 166.

present before man comes to discover the essentially religious character of interpersonal relations. In the same way, prayer is already present before man comes to discover that every potentiality, need, or request is, in fact, a prayer, because it is a call for God and can be nothing else.

Therefore, prayer is not some kind of vague longing for infinity, as it is sometimes represented in discussions of religion, religiousness, and the *"desiderium naturale."* Prayer is the human need for humanity and human things, and this is in fact—and can only be—a need for God, because all things come from him and all things re-present him.[122] This, of course, does not exclude the possibility that human desire can be vague and directionless, but it is not here—that is, in contrast to a specific human need—that the religiousness of man is founded.

Explicit, conscious, and reflexive prayer rests on the real and fundamental situation of man with respect to God. Prayer is the thematic expression and unfolding of man's situation, just as thematic religion is the unfolding of the actual religious situation. And just as religion becomes a reality for man as man only in thematic presentation and in sacrament, so, in the same way, man's need of God becomes a reality for man himself only in thematic prayer.

All the other questions about prayer that arise from ecclesial tradition and actual practice (keeping Thomas occupied for seventeen articles) must be approached in the light of this fundamental perspective.

The efficacy of prayer is not subject to doubt. Scripture bears witness that all who ask will receive (Mt. 7, 8), and the Church's tradition also shows a lively conviction that prayer cannot lack its effect. In the first place, however, the effect of prayer can be no other than that of the requesting, and this implies an imperfect causality: the expression of a need and of the willingness to receive. But in order to achieve the desired goal, more is required (of the one who asks or of someone else) than the mere expression of a need, which represents only one partial aspect of the total activity.

To the extent that prayer is directed to God, the stress is laid not on the acquiring or receiving as such, but on acquiring or receiving *from God.*

In this sense, all prayer is really heard and, in a certain sense,

[122] Note, too, *Summa Theologiae* I, 44, 4 ad 3; etc.

it bears its hearing within itself. For by prayer man directs himself to God,[123] acknowledging him as the one from whom all good comes,[124] irrespective of the particular human and earthly causes and of the effort—sometimes great effort—that one must make and by which the good gifts of God are made visible and realized.

Thus, the efficacy of prayer is always related to the actual relation of the man to God which is expressed and established in prayer. Consequently, it relates to all that man actually does receive, insofar as it is included or becomes included in prayer, for it is in this way that receiving-*from-God* is acknowledged and expressed.

It is understandable, then, why tradition, in addition to saying that all prayer is heard, can also say that one does not always receive what one prays for. For prayer is not a cause that comes into play when other causes have failed, but it is a reference to God as the ultimate cause and origin of all that we actually do receive.

Prayer for others is efficacious, but does not always receive the answer requested. As with any kind of prayer, here again there is no question of some kind of mysterious efficacy brought about in mysterious ways that are beyond us. Such a view is all too human, and does not attain to the actual sense and significance of prayer. No prayer whatever, even prayer for others, can make God change his mind. However, when we state that God cannot change and that his plans cannot be amended by our prayer, we must again be aware of an all too human view of God which pictures him as immovable and unfeeling as a rock. The contrary is the case. Our own compassion for a fellowman can come only from God, from his infinitely greater care for every individual man, in which he not only leaves men free, but gives them the freedom to turn against him.

Therefore the tradition declares not only that the efficacy of our prayer for others depends on the disposition of the others themselves, but also, and even more emphatically, that it is an essential demand of love of neighbor that we do pray for others.[125] For in this way we express God's own care and love for the other, which, in the other, is able to do more than our human mind can imagine or conceive.

[123] Cf. *ibid.*, II–II, 83, 1 ad 2.
[124] Cf. Jas. 1, 17; St. Thomas, *loc. cit.*, II–II, 83, 4c and ad 1.
[125] Cf. *ibid.*, II–II, 83, 7; 25, 9.

Let us add a final word on one of the many questions arising in Thomas' articles on prayer, because this question is a most striking combination of theological acuteness and sober judgment.

Must prayer be lengthy? This is the question. Scripture says that we must pray without ceasing.[126] We recognize Thomas' fundamental principle when he says that in a certain sense one must pray always and actually does pray always. But we usually understand the word "prayer" as reflexive or thematic prayer, and this kind of prayer cannot continue indefinitely for the simple reason that we have other things to do. Then with Augustine, Thomas further remarks that the individual, and the community too, must take into account the limits of human endurance in any *"fervor desiderii"* when deciding on the length of prayer, and not go on and on *"ita quod sine taedio durare non possit"* ("until one becomes bored stiff").

Religion and prayer, then, are seen to be reality and the manifestation of something that may no longer be regarded as service, but as friendship. In the sacramental life of the Church, and particularly in the eucharist, this friendship, which embraces every facet of the whole of human life, assumes an exemplary form.

[126] Lk. 18, 1; 1 Thess. 5, 17; cf. St. Thomas *loc. cit.*, II-II, 83, 14.

Index